LESSON QUESTION (What Question Are We Trying to Answer?) & LESSON NICKNAME	**WHAT DID WE FIGURE OUT?** Which parts of what we figured out (if any) can help us with our model? (Highlight them.)	**BASED ON OUR PROGRESS THIS LESSON, HOW CAN WE ADD TO OR REVISE OUR FINAL MODEL?** How should we represent our ideas in our model? (Use pictures, words, or symbols)	**CAN WE REPRESENT OUR NEW IDEAS USING EXISTING MODELS?** Can we use what we've learned before to represent our new ideas?
L2			
L3			

AF854469

LESSON QUESTION (What Question Are We Trying to Answer?) & LESSON NICKNAME	**WHAT DID WE FIGURE OUT?** Which parts of what we figured out (if any) can help us with our model? (Highlight them!)	**BASED ON OUR PROGRESS THIS LESSON, HOW CAN WE ADD TO OR REVISE OUR FINAL MODEL?** How should we represent our ideas in our model? (Use pictures, words, or symbols)	**CAN WE REPRESENT OUR NEW IDEAS USING EXISTING MODELS?** Can we use what we've learned before to represent our new ideas?
L4			
L5			

inquiryHub Chemistry

A 3-Dimensional High School Chemistry Curriculum

Student Materials
Unit 2 Fuels

Version 1.0

These materials were developed by the inquiryHub Research Practice Partnership, including educators from CO and IL, Northwestern University and the University of Colorado Boulder, with support from the Hewlett Foundation, the Spencer Foundation, the National Science Foundation, and Denver Public Schools.

Course Authors

Course Leadership Team
PI: William R. Penuel, CU Boulder
Co-PI: Tamara Sumner, CU Boulder
Co-PI: Brian J. Reiser, Northwestern University
Co-PI: Susan Olezene, Denver Public Schools
Melissa Campanella, CU Boulder
Kate Henson, CU Boulder
Kristin Rademaker, consultant
William Reed, Chicago Public Schools
Douglas Watkins, Denver Public Schools
Kerri Wingert, CU Boulder

Research Team
Anna-Ruth Allen, CU Boulder
Stefanie Christensen, Consultant
Yasko Endo, CU Boulder
Madison Hammer, CU Boulder
Kate Henson, CU Boulder
Katie Van Horne, CU Boulder
Kerri Wingert, CU Boulder

Fuels
Lead: **Melissa Campanella,** CU Boulder
Michelle Blue, Denver Public Schools
Amanda Dolph, Caddo (Louisiana) Parish Schools
Holly Hereau, BSCS Science Learning
Ari Leventhal, Denver Public Schools
Anna Noble, CU Boulder
Brian J. Reiser, Northwestern University
Will Reed, Chicago Public Schools
Kirk Sang, Denver Public Schools
Nicole Vick, BSCS Science Learning
Aliza Zivic, Northwestern University

Pilot Teachers

Trish Arnold
Patrick Aubin
Michelle Blue
Grace Boersma
Vince Bolton
Jaso Bowdre
Lisa Bradley
Samantha DeMatteo
Amanda Dolph
Tj Donahue
Ethan Dusto
Sarah Evans
Christine Garcia
Jeremy Gilbert
David Grossman
Jacobo Guevara Mendoza
Alex Hildebrand
Andrea Jacobberger
Katie Kleeman
Praveena Kotta
Lisa Kramer
Emily Larson
Noah Lederach
Roberto Maurera
Kristin Mayer
Rachel Patton
Will Reed
John Speier
Scotty Stalp
Derek Stoneman
Nicole Vick
Jessica Youmans
Michelle Zhang

inquiryHub Chemistry

A 3-Dimensional High School Chemistry Curriculum

Student Materials - Printable Resource
Unit 2 Fuels

The **Student Materials** printable resource for Unit 2: Fuels of inquiryHub Chemistry contains the following materials for each lesson as applicable:

- Student Incremental Modeling Trackers
- Student Activity Sheets
- Supplemental Materials

Unit Question: Why do cars run on gasoline and not rocket fuel?

NAME: ______________________

PERIOD: ____________ DATE: ____________

INCREMENTAL MODELING TRACKER (IMT) STUDENTS - Fuels

Throughout our investigation of fuels, we will be using the Incremental Modeling Tracker (IMT) as a thinking tool to help you keep track of and figure out important discoveries. You will refer back to the IMT regularly to help prioritize ideas and revise or build on our models. We will also use the IMT as a way to think with others about what is important in our models.

DRIVING QUESTION: Why do cars run on gasoline and not rocket fuel?

LESSON QUESTION (What Question Are We Trying to Answer?) & LESSON NICKNAME	**WHAT DID WE FIGURE OUT?** Which parts of what we figured out (if any) can help us with our model? (Highlight them!)	**BASED ON OUR PROGRESS THIS LESSON, HOW CAN WE ADD TO OR REVISE OUR FINAL MODEL?** How should we represent our ideas in our model? (Use pictures, words, or symbols)	**CAN WE REPRESENT OUR NEW IDEAS USING EXISTING MODELS?** Can we use what we've learned before to represent our new ideas?
L1			

inquiryHub Chemistry

A 3-Dimensional High School Chemistry Curriculum

Student Materials - Printable Resource
Unit 2 Fuels

The **Student Materials** printable resource for Unit 2: Fuels of inquiryHub Chemistry contains the following materials for each lesson as applicable:

- Student Incremental Modeling Trackers
- Student Activity Sheets
- Supplemental Materials

Unit Question: Why do cars run on gasoline and not rocket fuel?

NAME: ____________________

PERIOD: ____________ DATE: ____________

INCREMENTAL MODELING TRACKER (IMT) STUDENTS - Fuels

Throughout our investigation of fuels, we will be using the Incremental Modeling Tracker (IMT) as a thinking tool to help you keep track of and figure out important discoveries. You will refer back to the IMT regularly to help prioritize ideas and revise or build on our models. We will also use the IMT as a way to think with others about what is important in our models.

DRIVING QUESTION: Why do cars run on gasoline and not rocket fuel?

LESSON QUESTION (What Question Are We Trying to Answer?) & LESSON NICKNAME	**WHAT DID WE FIGURE OUT?** Which parts of what we figured out (if any) can help us with our model? (Highlight them!)	**BASED ON OUR PROGRESS THIS LESSON, HOW CAN WE ADD TO OR REVISE OUR FINAL MODEL?** How should we represent our ideas in our model? (Use pictures, words, or symbols)	**CAN WE REPRESENT OUR NEW IDEAS USING EXISTING MODELS?** Can we use what we've learned before to represent our new ideas?
L1			

These materials were developed by the inquiryHub RPP, including educators from CO and IL, Northwestern University and the University of Colorado Boulder, with support from the Hewlett Foundation, the Spencer Foundation, the National Science Foundation, and Denver Public Schools.

LESSON QUESTION (What Question Are We Try ng to Answer?) & LESSON NICKNAME	**WHAT DID WE FIGURE OUT?** Which parts of what we figured out (if any) can help us with our model? (Highlight them!)	**BASED ON OUR PROGRESS THIS LESSON, HOW CAN WE ADD TO OR REVISE OUR FINAL MODEL?** How should we represent our ideas in our model? (Use pictures, words, or symbols)	**CAN WE REPRESENT OUR NEW IDEAS USING EXISTING MODELS?** Can we use what we've learned before to represent our new ideas?
L6			
L7			

LESSON QUESTION (What Question Are We Trying to Answer?) & LESSON NICKNAME	**WHAT DID WE FIGURE OUT?** Which parts of what we figured out (if any) can help us with our model? (Highlight them!)	**BASED ON OUR PROGRESS THIS LESSON, HOW CAN WE ADD TO OR REVISE OUR FINAL MODEL?** How should we represent our ideas in our model? (Use pictures, words, or symbols)	**CAN WE REPRESENT OUR NEW IDEAS USING EXISTING MODELS?** Can we use what we've learned before to represent our new ideas?
L8			
L9			

LESSON QUESTION (What Question Are We Trying to Answer?) & LESSON NICKNAME	**WHAT DID WE FIGURE OUT?** Which parts of what we figured out (if any) can help us with our model? (Highlight them!)	**BASED ON OUR PROGRESS THIS LESSON, HOW CAN WE ADD TO OR REVISE OUR FINAL MODEL?** How should we represent our ideas in our model? (Use pictures, words, or symbols)	**CAN WE REPRESENT OUR NEW IDEAS USING EXISTING MODELS?** Can we use what we've learned before to represent our new ideas?
L10			
L11			

LESSON QUESTION (What Question Are We Trying to Answer?) & LESSON NICKNAME	**WHAT DID WE FIGURE OUT?** Which parts of what we figured out (if any) can help us with our model? (Highlight them!)	**BASED ON OUR PROGRESS THIS LESSON, HOW CAN WE ADD TO OR REVISE OUR FINAL MODEL?** How should we represent our ideas in our model? (Use pictures, words, or symbols)	**CAN WE REPRESENT OUR NEW IDEAS USING EXISTING MODELS?** Can we use what we've learned before to represent our new ideas?
L12			
L13			

LESSON QUESTION (What Question Are We Trying to Answer?) & LESSON NICKNAME	**WHAT DID WE FIGURE OUT?** Which parts of what we figured out (if any) can help us with our model? (Highlight them!)	**BASED ON OUR PROGRESS THIS LESSON, HOW CAN WE ADD TO OR REVISE OUR FINAL MODEL?** How should we represent our ideas in our model? (Use pictures, words, or symbols)	**CAN WE REPRESENT OUR NEW IDEAS USING EXISTING MODELS?** Can we use what we've learned before to represent our new ideas?
L10			
L11			

LESSON QUESTION (What Question Are We Trying to Answer?) & LESSON NICKNAME	**WHAT DID WE FIGURE OUT?** Which parts of what we figured out (if any) can help us with our model? (Highlight them!)	**BASED ON OUR PROGRESS THIS LESSON, HOW CAN WE ADD TO OR REVISE OUR FINAL MODEL?** How should we represent our ideas in our model? (Use pictures, words, or symbols)	**CAN WE REPRESENT OUR NEW IDEAS USING EXISTING MODELS?** Can we use what we've learned before to represent our new ideas?
L12			
L13			

LESSON QUESTION (What Question Are We Trying to Answer?) & LESSON NICKNAME	**WHAT DID WE FIGURE OUT?** Which parts of what we figured out (if any) can help us with our model? (Highlight them!)	**BASED ON OUR PROGRESS THIS LESSON, HOW CAN WE ADD TO OR REVISE OUR FINAL MODEL?** How should we represent our ideas in our model? (Use pictures, words, or symbols)	**CAN WE REPRESENT OUR NEW IDEAS USING EXISTING MODELS?** Can we use what we've learned before to represent our new ideas?
L14			
L15			

NAME: ____________________

PERIOD: __________ DATE: __________

Lesson 1: Why do cars run on gasoline and not rocket fuel?

BLUE SPHERES VIDEO

As you watch the video presented in class, record your thinking.

1. What do you think is going on here?

2. What do you notice in the video that makes you say that?

3. What questions do you have about what is being shown?

4. Share your ideas with a partner or small group, make some connections. What does this video of the blue spheres have to do with us?

What did you notice that might connect to something in your life?	What did you notice that extends or pushes your thinking in new directions?	What did you notice that was challenging or confusing that leaves you with questions?

5. Demonstrate your thinking

Modeling Greenhouse gases (use the space below for brainstorming)

DATA

6. Examine *Figure ES-14: US Greenhouse Gas Emissions Allocated to Economic Sectors.*

What do you **notice**?	
What do you **think** about what you notice? ...	
What do you **wonder**? ?	

Record your ideas about what this graph is telling us:

7. Examine the ***NYT Interactive Map of Transportation Emissions*** in your area.

What do you **notice**?	
What do you **think** about what you notice? ...	

What do you **wonder?** ?	

Record your ideas about what this graph is telling us.

8. Examine *the bar graph of 2017 US CO_2 Emissions from Fossil Fuel combustion by Fuel Type and End-Use Sector.*

What do you **notice**?	
What do you **think** about what you notice? ...	
What do you **wonder**? ?	

Record your ideas about what this graph is telling us.

9. Examine the document: *Average C02 emissions by car type.*

What do you **notice**?	
What do you **think** about what you notice?	
What do you **wonder**?	

Record your ideas about how this connects to the other graphs and what we saw in the blue sphere video.

10. Examine the document *US transportation energy sources/fuels, 2019.*

What do you **notice**?	
What do you **think** about what you notice?	
What do you **wonder**?	

Record your ideas about how this connects to the other graphs and what we saw in the turquoise sphere video.

11. Examine the document: *Energy and CO_2 emissions per fuel type.*

What do you **notice**?	
What do you **think** about what you notice?	
What do you **wonder**? ?	

12. Describe the general pattern you see in the data for *Energy and CO_2 emissions per fuel type.*

13. What do all of these fuels have in common?

14. How are they different?

SYNTHESIZING / NEXT STEPS

15. What is a big takeaway from the data we looked at today that helps us think about what we could do to reduce CO_2 emissions from transportation?

16. What should we investigate next?

17. Describe how we could perform the investigation you suggested. (What materials or resources would we need and how would we use them to answer our question)

READING: LIQUID HYDROGEN - THE FUEL OF CHOICE FOR SPACE EXPLORATION

18. Read the article, Rocket Fuel, two times. The first time, focus on any similarities between how hydrogen is used in a rocket and how gasoline is used in a car. The second time, focus on any differences.

Similarities	Differences

BRAINSTORMING RELATED PHENOMENA

19. What might be examples of combustion/burning we've seen before (in or outside class) where the explosion or fire was more or less intense? Where burning resulted in more or less exhaust/smoke or other byproducts?

INITIAL MODELS - INDIVIDUAL

20. Use the template on the following page or use the space below to create an initial model that answers the question:

Both gasoline and hydrogen fuels burn or combust in car and rocket engines. If the process they undergo is the same, what might be going on that causes Hydrogen to produce so much more energy per gram than Gasoline?

In your model, do your best to explain following points for both Hydrogen and Gasoline:

- What is happening before, during, and after combustion that could explain this
- What is happening at both the particle and observable scales that could explain this

Model - Combustion of Gasoline Fuel in a Car

Observable Scale

before | during | after

Particle Scale

before | during | after

Image: pexels.com/ Mike

Additional Evidence/Explanation:

Model - Combustion of Hydrogen Fuel in a Rocket

Observable Scale

before | during | after

Particle Scale

before | during | after

Image: NASA (public domain)

Additional Evidence/Explanation:

INITIAL MODELS - SMALL GROUP

Share your individual models with a small group and work together to build a model together on a piece of chart paper.

GOING PUBLIC WITH OUR MODELS

21. Summarize each team's model and write out your thoughts and questions in the space below. This will help us build our Driving Question Board!

Summary of Model (How did this group explain why burning hydrogen would produce more energy per gram than burning gasoline?)	Similarities and Differences to your group's consensus model	Thoughts and Questions

CONSENSUS BUILDING DISCUSSION

22. What questions does this bring up for you? Look through these pages to help you think about questions you came up with at different times in this lesson, write each question and any news questions you have on an individual sticky note or notecard.

NAME: ______________________

PERIOD: ____________ DATE: ____________

Lesson 1: Emissions by Sector

Figure ES-14: U.S. Greenhouse Gas Emissions Allocated to Economic Sectors (MMT CO_2 Eq.)

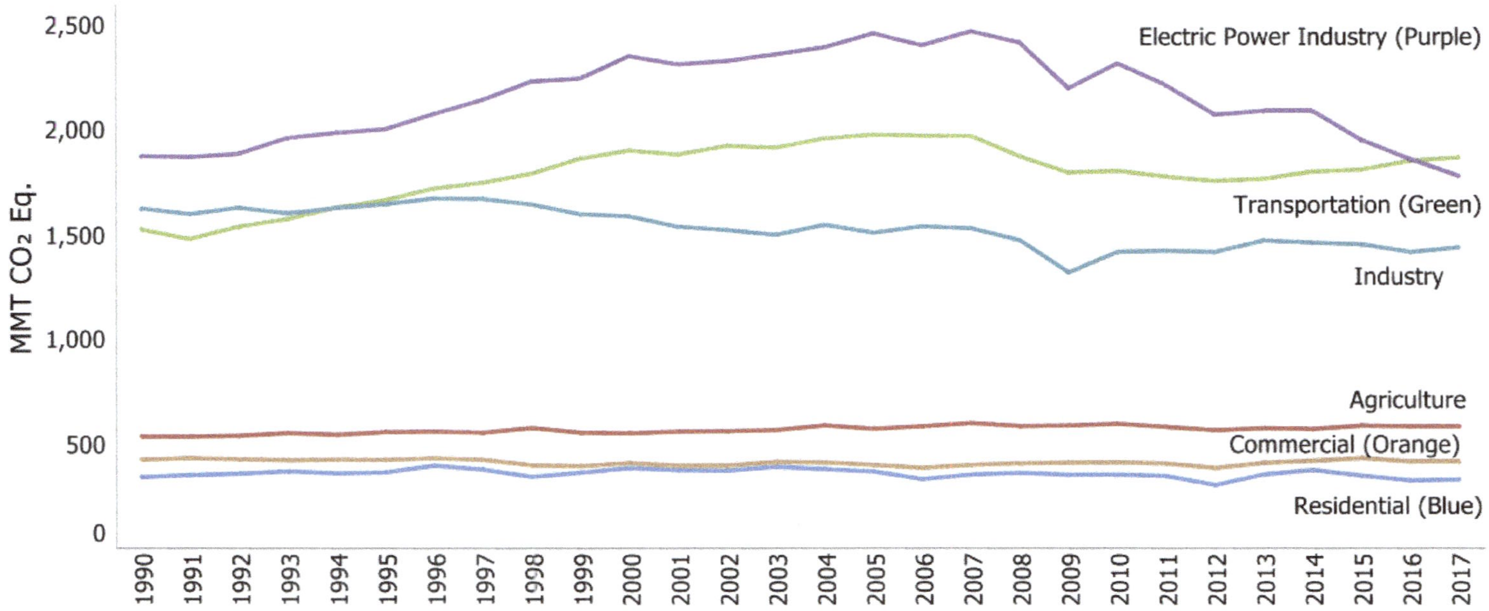

Source: Environmental Protection Agency

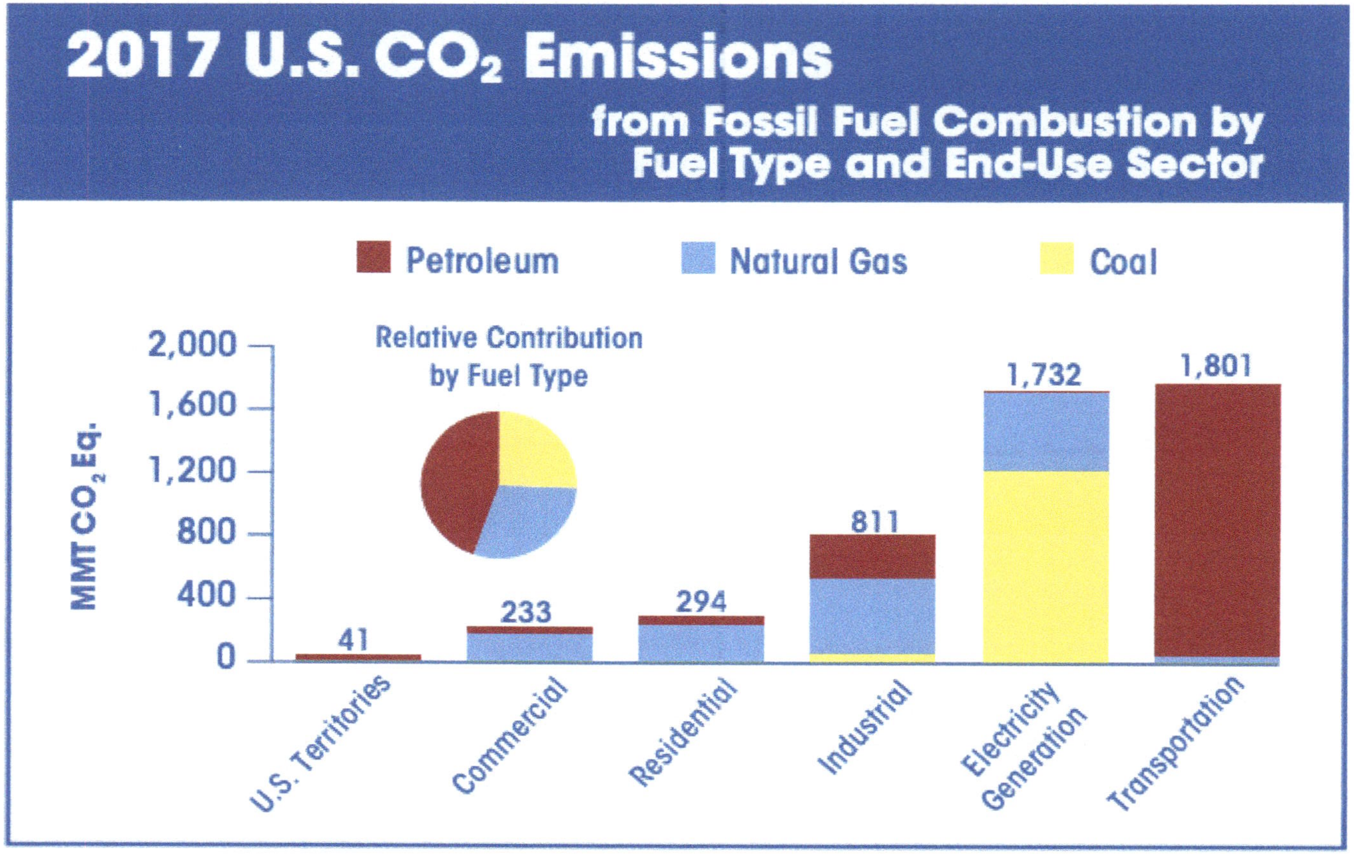

Note: Electricity generation also includes emissions of less than 0.5 Tg CO_2 Eq. from geothermal-based electricity generation.

Data and Image Source: Environmental Protection Agency (EPA)

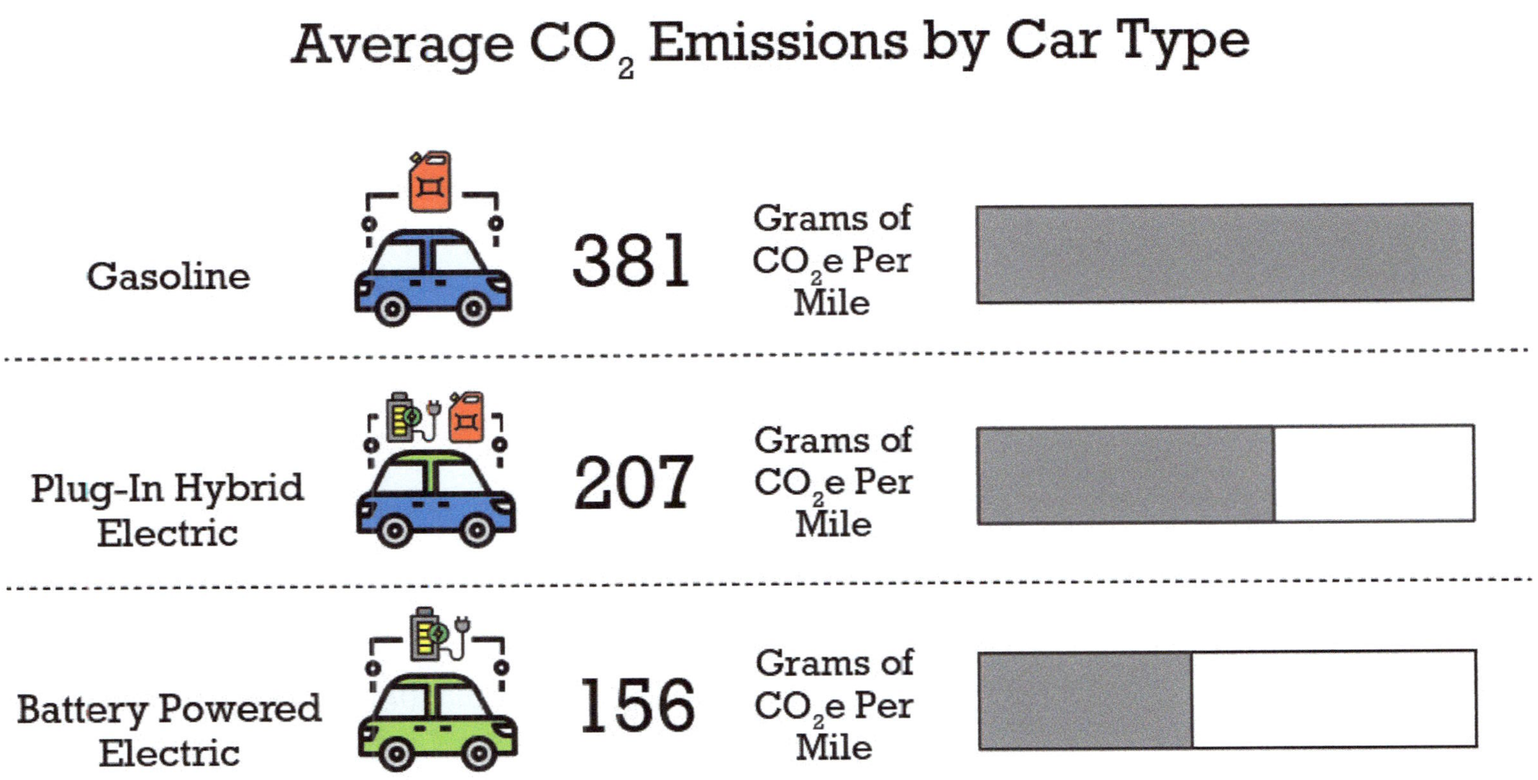

Image: Graph created by MH for inquiryHub, Data from CleanTechnica

U.S. transportation energy sources/fuels, 2019 1

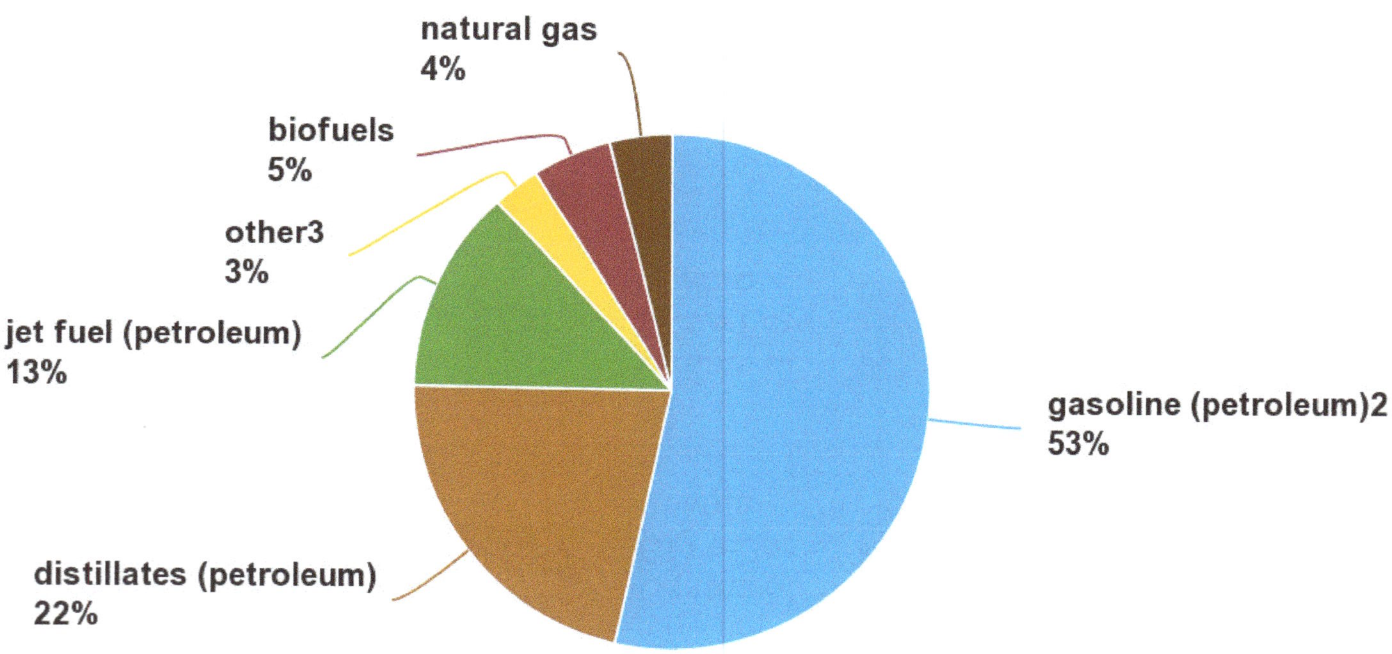

Note: Sum of individual components may not equal 100% because of independent rounding.

Source: U.S. Energy Information Administration, *Monthly Energy Review*, Tables 2.5, 3.8c, and 10.2b, May 2020, preliminary data

Fuel	Energy from Combustion (MJ/kg of fuel)	Emissions from Combustion (lbs CO_2 /kg of fuel)
Ethanol	29.7	4.4
Gasoline	46.5	7.1
Diesel	45.8	6.8
Biodiesel	39.6	5.7
Natural Gas	55.0	5.3
Jet Fuel	44.6	6.9
Hydrogen	141.9	0

1 MJ (megajoule) is equal to 1000 kJ (kilojoules)

Fuel properties comparison data from: U.S. Department of Energy - Alternate Fuels Data Center

SOURCES

Benson, T. (Ed.). (2014, June 12). Combustion. Retrieved June 29, 2020, from https://www.grc.nasa.gov/WWW/K-12/rocket/combst1.html

Braeunig, R. (2008). Rocket Propellants. Retrieved June 29, 2020, from http://www.braeunig.us/space/propel.htm

Dunbar, B. (2010, July 29). Liquid Hydrogen--the Fuel of Choice for Space Exploration. Retrieved June 29, 2020, from https://www.nasa.gov/topics/technology/hydrogen/hydrogen_fuel_of_choice.html

Perry, B. (2016, April 21). Chemical reactions – Rocketology: NASA's Space Launch System. Retrieved June 29, 2020, from https://blogs.nasa.gov/Rocketology/tag/chemical-reactions/

The Science Guys. (2002, November). Retrieved June 29, 2020, from https://www.uu.edu/dept/physics/scienceguys/2002Nov.cfm

NAME: ______________________

PERIOD: ____________ DATE: ____________

Lesson 1: Fuels Source

Fuel	Energy from Combustion (MJ/kg of fuel)	Emissions from Combustion (lbs CO_2 /kg of fuel)
Ethanol	29.7	4.4
Gasoline	46.5	7.1
Diesel	45.8	6.8
Biodiesel	39.6	5.7
Natural Gas	55.0	5.3
Jet Fuel	44.6	6.9
Hydrogen	141.9	0

1 MJ (Megajoule) is equal to 1000 kJ (kiloJoules)

Fuel properties comparison data from: U.S. Department of Energy - Alternative Fuels Data Center

These materials were developed by the inquiryHub RPP, including educators from CO and IL, Northwestern University and the University of Colorado Boulder, with support from the Hewlett Foundation, the Spencer Foundation, the National Science Foundation, and Denver Public Schools.

SOURCES

Dickerson, R., & Geis, I. (1979). Chemistry, matter, and the universe: An integrated approach to general chemistry. In Chemistry, matter, and the universe: An integrated approach to general chemistry. Chapter 12 (Chapter 12, p. 13). Menlo Park: Benjamin/Cummings. Retrieved March 13, 2021, from http://www.chem.ox.ac.uk/vrchemistry/energy/Page_13.htm

Owolabi, Rasheed & A, Osiyemi & Amosa, Mutiu & Elizabeth, Ojewumi. (2011). Biodiesel from Household/Restaurant Waste Cooking Oil (WCO). Journal of Chemical Engineering & Process Technology. 2. 1-4. 10.4172/2157-7048.1000112.

Quaschning, V. (2015, June). Specific carbon dioxide emissions of various fuels. Retrieved March 14, 2021, from https://www.volker-quaschning.de/datserv/CO2-spez/index_e.php

World Nuclear Association. Heat Values of Various Fuels. Retrieved March 14, 2021, from https://www.world-nuclear.org/information-library/facts-and-figures/heat-values-of-various-fuels.aspx

NAME: ____________________

PERIOD: __________ DATE: __________

Lesson 1: Reading - Liquid Hydrogen: The Fuel of Choice for Space Exploration

Image source: NASA

Liquid hydrogen is the signature fuel of the American space program and is used by other countries in the business of launching satellites. Taming liquid hydrogen is one of the significant technical achievements of twentieth century American rocketry. Jet engines are air breathers. They take in air (which contains oxygen needed for **combustion**), mix it with fuel, burn it to increase the pressure, and exhaust the spent gases out the back at a high rate of speed. This high-speed ejection of mass propels the plane forward. Rockets do almost the same thing but unlike jets, they carry their own oxygen along with them. Cars don't need to carry oxygen for their fuel burning action, the air is filled with oxygen. Cars, like jets, take in air to burn with fuel which increases pressure so they can use that pressure to push pistons which in turn rotate the wheels of the car.

During combustion, new chemical substances are created from the fuel and the oxidizer. These substances are called **exhaust**. Most of the exhaust comes from chemical combinations of the fuel and oxygen. When a hydrogen-carbon-based fuel (like gasoline) burns, the exhaust includes water (2 hydrogens + 1 oxygen) and carbon dioxide (1 carbon + 2 oxygens).

During the combustion process, as the fuel and oxidizer are turned into exhaust products, heat is generated. Interestingly, some source of heat is usually necessary to initiate combustion. Gasoline and air are both present in your automobile fuel tank; but combustion does not occur because there is no source of heat. Since heat is both required to start combustion and is itself a **product** of combustion, we can see why combustion continues to happen: heat starts the combustion, combustion produces heat which keeps the reaction going. Once combustion gets started, we don't have to provide the heat source because the heat of combustion continues to activate the reactants. Much like we don't have to keep lighting a campfire, it just keeps burning.

Despite criticism and early technical failures, the taming of liquid hydrogen proved to be one of NASA's most significant technical accomplishments. Hydrogen -- a light and extremely powerful rocket propellant -- has the lowest molecular weight of any known substance and burns with extreme intensity (5,500°F). In combination with an oxidizer such as liquid oxygen, liquid hydrogen yields the highest specific impulse, or efficiency in relation to the amount of propellant consumed, of any known rocket propellant. The chemical energy stored in the fuel allows the rocket to go thousands of miles in just a couple of minutes. Hydrogen is an amazing fuel.

The high efficiency of liquid oxygen/liquid hydrogen makes it ideal for powering rockets. Liquid hydrogen delivers a specific impulse (a measure of how effectively a rocket uses propellant or a jet engine uses fuel) about 30%-40% higher than most other fuels. Gasoline has been a successful fuel for a long time, helping power our cars and other engines. Whether it's your personal vehicle or a behemoth launch vehicle like Space Shuttle, the basics are the same. Combustion (burning something) releases energy, which makes things go. Start with fuel (something to burn) and an oxidizer (something to make it burn) and now you've got propellant (something to give a push) and the byproduct. Give the fuel and oxygen a spark and energy is released, along with some byproducts. For cars the byproduct is greenhouse gasses that can contribute to climate change. The hydrogen combustion reaction is sometimes considered "green" because it does not release any carbon dioxide.

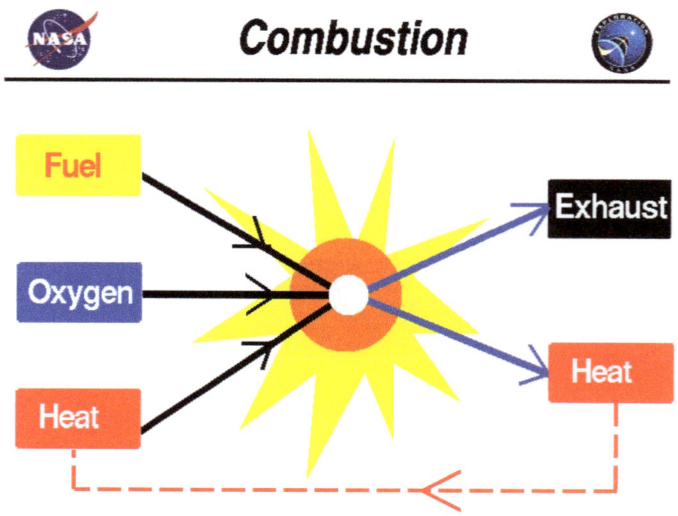

Image source: NASA

This "green" reaction releases massive amounts of energy along with superheated water (steam). The hydrogen-oxygen reaction inside a rocket engine generates tremendous heat, causing the water vapor to expand and exit the engine nozzles at speeds of 10,000 miles per hour! All that fast-moving steam creates the thrust that propels the rocket from Earth.

SOURCES

Benson, T. (Ed.). (2014, June 12). Combustion. Retrieved June 29, 2020, from https://www.grc.nasa.gov/WWW/K-12/rocket/combst1.html

Braeunig, R. (2008). Rocket Propellants. Retrieved June 29, 2020, from http://www.braeunig.us/space/propel.htm

Dunbar, B. (2010, July 29). Liquid Hydrogen--the Fuel of Choice for Space Exploration. Retrieved June 29, 2020, from https://www.nasa.gov/topics/technology/hydrogen/hydrogen_fuel_of_choice.html

Perry, B. (2016, April 21). Chemical reactions – Rocketology: NASA's Space Launch System. Retrieved June 29, 2020, from https://blogs.nasa.gov/Rocketology/tag/chemical-reactions/

The Science Guys. (2002, November). Retrieved June 29, 2020, from https://www.uu.edu/dept/physics/scienceguys/2002Nov.cfm

NAME: ______________________

PERIOD: __________ DATE: __________

Lesson 2: What is it about burning a fuel that makes a car or rocket move?

LOOKING BACK

1. What did we figure out last time about the fuels that are used to make vehicles like cars and rockets move?

LOOKING FORWARD

2. Use writing, drawing or a combination of these to explain the **connection(s) between burning and energy?**

Image: BioRender.com

3. Use writing, drawing or a combination of these to explain the connections between **energy, matter,** and **movement** (*both at the particle scale, and on a scale that we can see?*)

DESIGN CHALLENGE

Can we manipulate temperature/particle scale movement to make the syringe plunger *move* at the observable scale?

4. Plan your design using these materials:

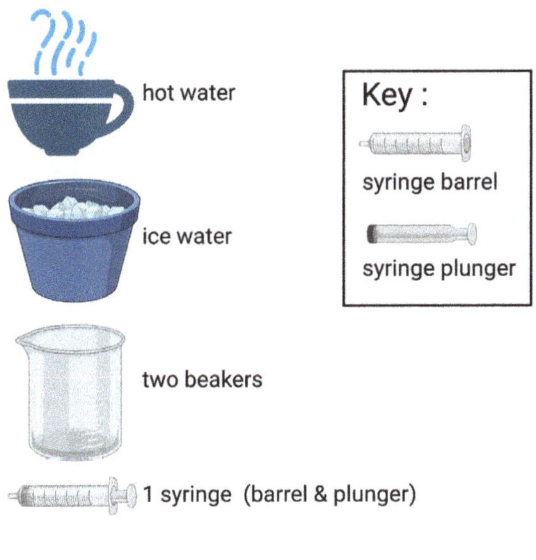

Images from: BioRender.com

(Draw / label a diagram of your design)

5. What **data** do you need to collect and **how much**?

6. What **accuracy** is needed? How will you get **reliable measurements?**

7. How will you **collaborate** to collect this data and share it with other teams?

Safety considerations:

8. Build and test your design. Record your data and notes below.

Test 1
Data:
Notes:

9. Explain any **refinements** to your design and your rationale. **What did/would you change** to make your design work better **and why?**

Test 2 (if time)
Data:
Notes:

10. Go Public. Share your results with another team. Listen to their representative share their results. **What are you thinking about now that you weren't before?**

MODELING THE SYRINGE SYSTEM

11. In the space below, create an initial model of what you think is happening with the syringe system. (You may use a separate sheet of paper if you would rather not use the template below)

What caused the syringe plunger to move when surrounded by hot water?
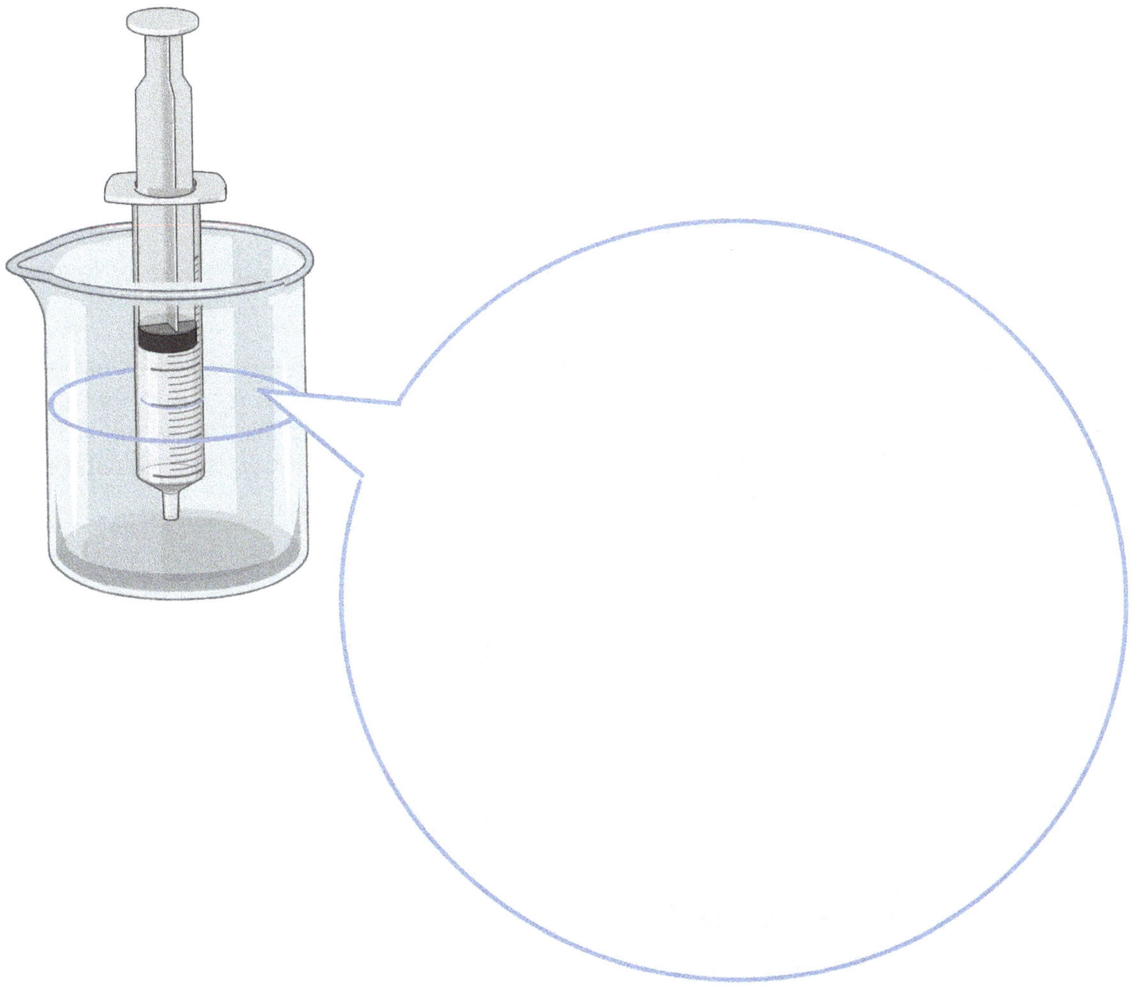 Image from: BioRender.com

12. Share your models with your small group to identify similarities and differences between your initial models. If there are ideas you would like to incorporate into future models, list them below.

BUILDING CONSENSUS

Add to or change your model as needed during the whole class consensus discussion. Make note of new questions this brings up for you.

13. Read this definition of the law of conservation of energy from NASA's Glenn Research Center.

Conservation of Energy - Within some problem domain, the amount of energy remains constant and energy is neither created nor destroyed. Energy can be converted from one form to another (potential energy can be converted to kinetic energy) but the total energy within the domain remains fixed.

Record your ideas about the following statements below:

→ If energy cannot be created or destroyed, just converted/transformed, *where is the energy coming from* when a fuel is burned in a car or rocket engine?

→ If energy is released when a fuel is burned, where might that energy have been *before*? In what *form(s)* might it have existed?

SLOW MOTION VIDEO - FOUR STROKE ENGINE

14. Take notes and/or draw particle models showing your observations regarding ENERGY, MATTER, and MOVEMENT for each stroke of the engine.

Stroke #1 - INTAKE 1 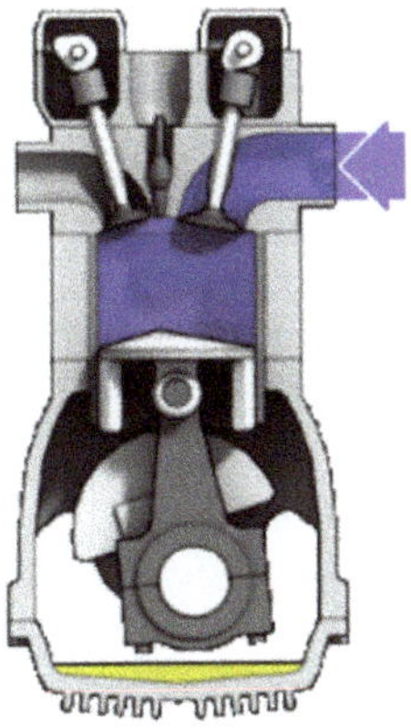 Image Credit: WikiMedia Commons/ Zephyris / CC BY-SA		
Stroke #2 - COMPRESSION 2  Image Credit: WikiMedia Commons/ Zephyris / CC BY-SA		

<u>Stroke #3 - POWER</u> 3 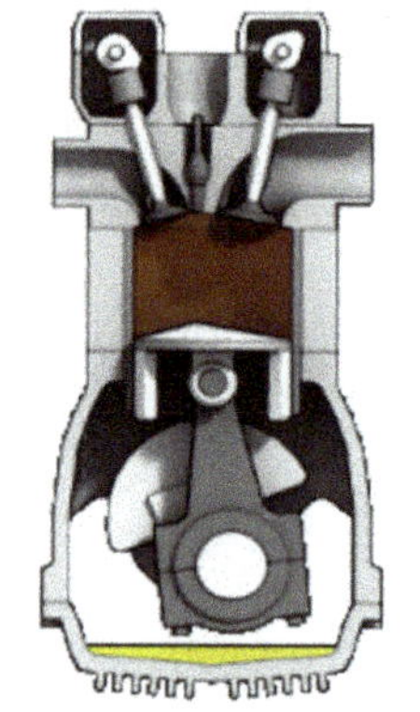 Image Credit: WikiMedia Commons / Zephyris / CC BY-SA		
<u>Stroke #4 - EXHAUST</u> 4 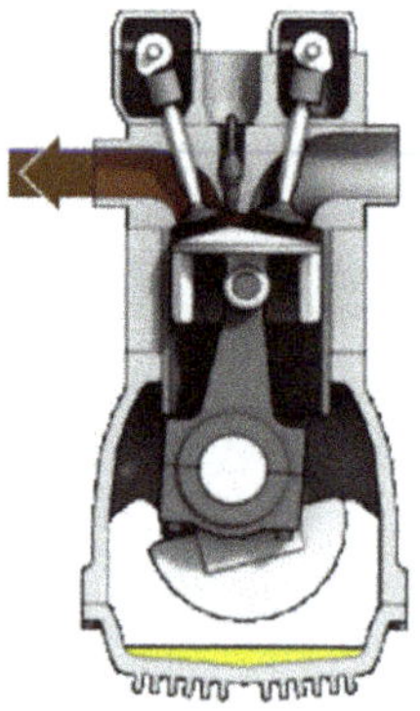 Image Credit: WikiMedia Commons / Zephyris / CC BY-SA		

How is there so much kinetic energy in by stroke #3? Where might the energy have come from (within or outside the fuel system)? What *form of energy* might it have converted from?

GO PUBLIC

Explain using words and/or drawings:

15. How is the kinetic energy in the system changing with each stroke of the combustion engine?

16. Where is the energy coming from when the kinetic energy of the system increases?

17. What energy transfers might be happening to account for these changes?
(in what other *forms* does the energy exist in the fuel/engine system?)

REPRESENTING ENERGY IN THE ENGINE

18. Next use the following energy diagrams to help you track how energy is changing in each stroke of the engine, where energy is coming from/going to, and how it is conserved.

Stroke #1- INTAKE 1 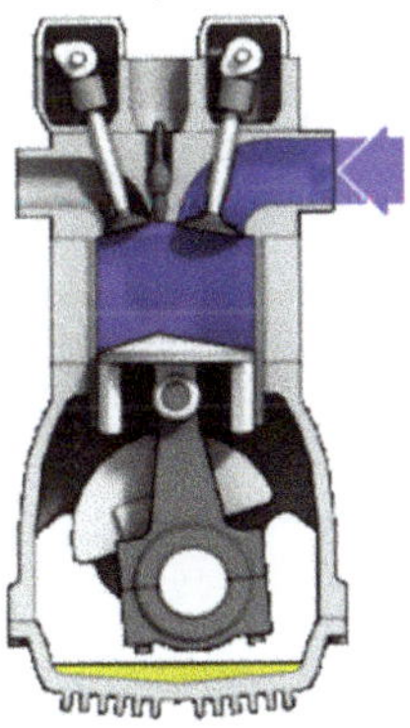 Images from: WikiMedia Commons/ Zephyris / CC BY-SA	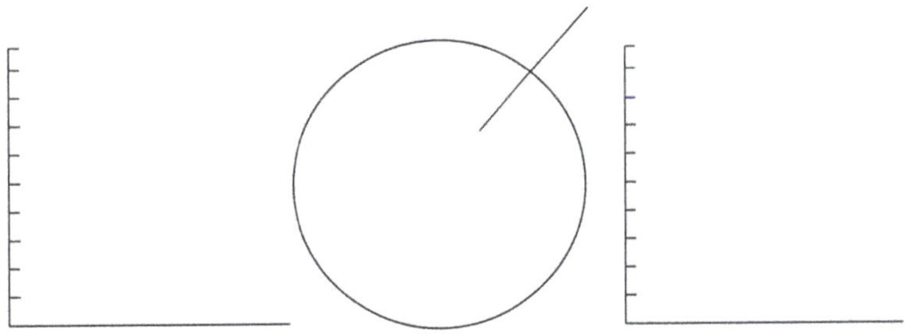

<u>Stroke #2 - COMPRESSION</u> 2 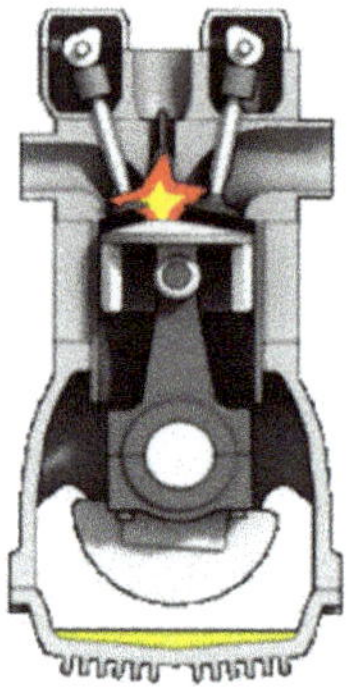 Images from: WikiMedia Commons/ Zephyris / CC BY-SA	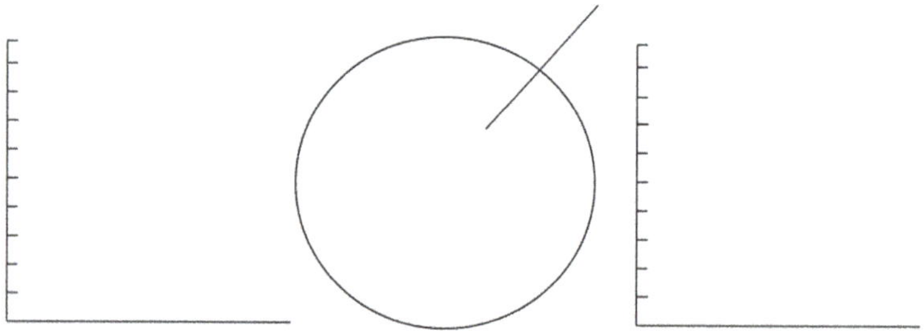
<u>Stroke #3 - POWER</u> 3 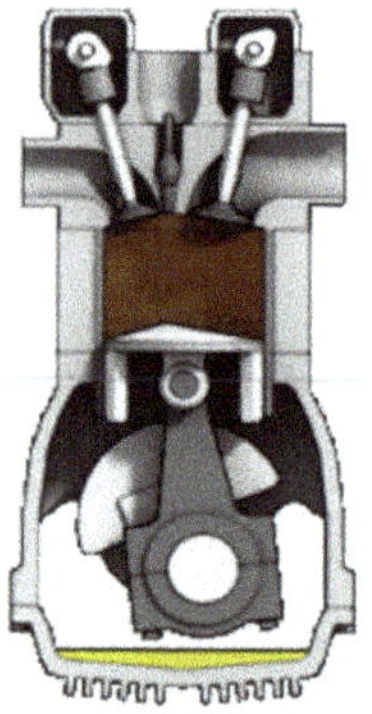 Image from: WikiMedia Commons/ Zephyris / CC BY-SA	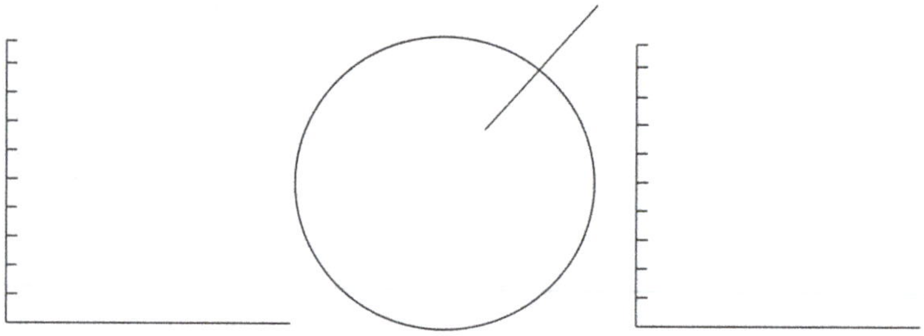
<u>Stroke #4 - EXHAUST</u> 4 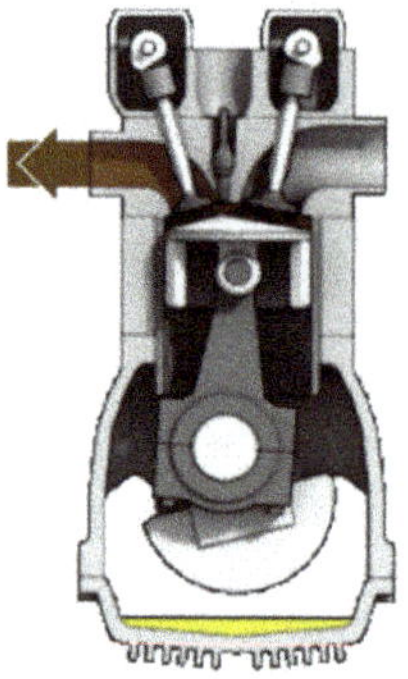 Image from: WikiMedia Commons/ Zephyris / CC BY-SA	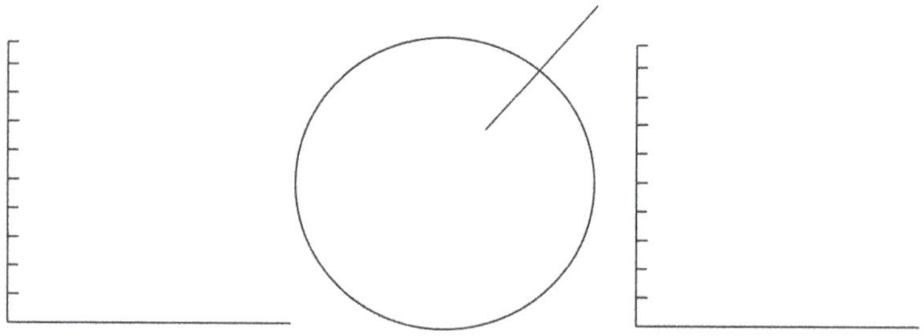

NEXT STEPS - HOW CAN WE RESOLVE THIS?

19. What do we need to do to figure out where the energy comes from when a fuel is burned?

20. How should we investigate this?

21. What evidence would we need to gather?

22. How might answering this question help us figure out what it is about hydrogen that makes the "explosion" so powerful?

NAME: ____________________

PERIOD: __________ DATE: __________

Lesson 3: What is burning and how does it give us energy?

INDIVIDUAL:

1. Use the space below to draw and explain how heating a syringe helped us understand how an engine works when fuel is burned.

2. What questions did the class raise at the end of the last lesson that we thought would be important to answer?

3. Use the space below to record your ideas for how we can investigate what the "smoke" that we saw leave the engine might be made of. When you finish recording them here, share your ideas with a partner.

4. Record your observations and questions as we investigate molecules that make up different types of fuel.

NOTICE	WONDER

WITH YOUR GROUP:

5. How you plan to carry out this investigation - for example:

a. What variables do you plan to measure?

b. How and when do you plan to measure these variables?

c. Sketch the experimental set up below - be sure to "label" what is in each dish, and what has been done to each dish.

d. Use the space below to create an organized data table to collect your measurements and observations over time.

e. Check with your teacher prior to lighting your ethanol: Teacher Initials ______________

INDIVIDUAL PREDICTIONS:

6. The class is attempting to find evidence to support an explanation of what kind of chemical process burning is. Make predictions about what your data would look like in each of the following situations:

If it is true that burning can be explained by Evaporation alone:

	Dish of BTB under container that didn't burn the ethanol	Dish of BTB under container that did burn the ethanol
Will BTB change color?	Yes / No	Yes / No
Explain why the outcomes you circled above would support the statement at the top of the table.		

If it is true that burning can be explained by combustion:

	Dish of BTB under container that didn't burn the ethanol	Dish of BTB under container that did burn the ethanol
Will BTB change color?	Yes / No	Yes / No
Explain why the outcomes you circled above would support the statement at the top of the table.		

INDIVIDUALLY:

7. In the space below, use phrases and pictures to record your observations for how both water and ethanol interact with the different surfaces you investigate. You will use these make inferences about what you observe happening under the container

Distilled water interacting with different surfaces	**Unknown liquid** droplets interacting with the sides of the glass container	**Ethanol** interacting with different surfaces

8. Describe the data you collected during the investigation with regard to:

a. BTB color change in the container where you ignited the ethanol and in the container where ethanol was not ignited.

b. Describe the characteristics of the drops that collected on the inside of the glass container where we burned the ethanol. Did the characteristics seem to be more like water or more like ethanol?

MAKING SENSE

9. Does the data collected above serve as evidence to support the idea that burning is one of the processes we set out to examine?

a. Does the data collected regarding the BTB color change support the idea that burning ethanol is combustion or evaporation? Explain why.

b. Does the data collected regarding the characteristics of the drops of liquid support the idea that burning ethanol is combustion or evaporation? Explain why.

10. Use the Gotta-Have-it checklist we developed in class to show how we can explain the process of burning fuel in each of the systems we talked about below. (It may help to copy the checklist below)

Record the Gotta-Have-It checklist here:		
Automobile Fuel (Engine)	**Food fuel (human body)**	**Wood Fuel (forest fire)**

WITH A PARTNER:

11. Construct a one-sentence definition of what burning is and record it here.

12. Use what you have learned in the course of this investigation to update the IMT (Incremental Model Tracker.)

NAME: ______________________________

PERIOD: ____________ DATE: ____________

Lesson 4: Could any chemical reaction be used to fuel a car? Do all chemical reactions "release" energy?

INDIVIDUAL:

1. What were the key observations (data) we collected when we burned ethanol that served as evidence to support the idea that burning is a chemical reaction?

2. What questions did the class raise at the end of the last lesson that we thought would be important to answer or test?

3. How would you use this information to revise our working model to explain how fuel makes an engine work (and therefore a car move)?

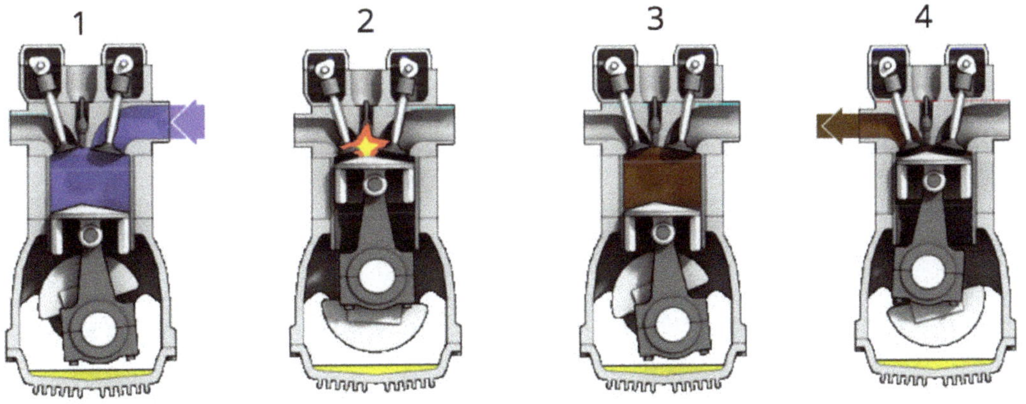

Image Source: WikiMedia Commons/ Zephyris / CC BY-SA

4. Do you have any experience with chemical reactions in your day to day activities? List them here and think about whether a reaction like you describe could power a car.

5. Record what you notice and any questions you have as the demonstration is taking place

Observations	Questions

6. Based on the observations above, does a chemical reaction take place when you mix these substances?

a. Which of these observations would you use to serve as data to support your claim?

b. Do you think a process like this could be used as energy for moving a car?

Apply what you learned from the class consensus model to the model of the engine below to explain whether or not a reaction must release heat to make a car go.

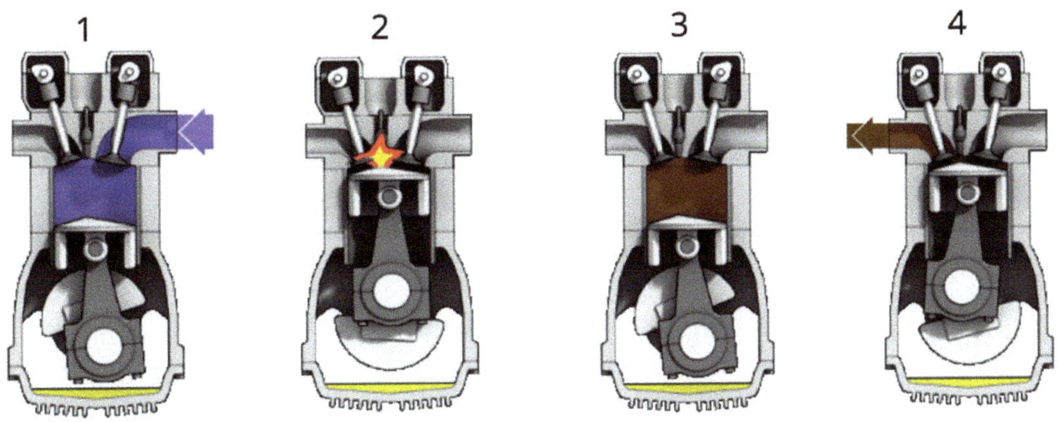

Image Source: WikiMedia Commons/ Zephyris / CC BY-SA

8. If we are looking to find alternative fuel sources, what are some criteria they should meet to power the engines we currently use?

9. What do we need to investigate next to help us understand what might meet these criteria?

NAME: ______________________

PERIOD: __________ DATE: __________

Lesson 5: Can we model what is happening when particles are rearranging to help us explain the changes we are seeing in energy?

RECAP

1. The last two labs helped us figure out that particles rearrange during chemical reactions and that this can result in an increase OR decrease in temperature. **What are your current ideas about how energy changes when bonds are made or broken? How might this connect to reactions heating up or cooling down?**

2. Today we will try to answer the question "Can we model what is happening when particles are rearranging to help us explain the changes we are seeing in energy?" **How might this help us figure out why hydrogen has such a high energy output compared to gasoline?**

PLANNING OUR INVESTIGATION

3. We have the following materials to use to represent the bonds between particles:

- Ball and stick models with plastic connectors
- Bar magnets
- Magnetic marbles
- Paper cut-outs of atoms

Review the chart with the pros and cons of using these materials to model bond making and breaking. **Which of these materials should we use today to model bond breaking and making and its relationship to energy (particle speed)? Why? What are the <u>limitations</u> of some of these materials?**

4. Your teacher has provided materials you can use in this next investigation. Use the space below to record your and your classmates' ideas about how to use these materials to physically model energy changes associated with bond making and bond breaking.

What in the physical model represents an atom?	
What in the physical model represents a bond?	
What action or movement represents bonds breaking?	
What action or movement represents bonds forming?	
What action or movement represents a change in energy (change in particle speed)?	
If the system in this investigation is the bond between two atoms, what would be outside of the system?	
How will we collect data on energy changes associated with bond breaking and making? It can be quantitative or qualitative.	

Draw a diagram of your setup here:

INVESTIGATION PART 1

5. Using your chosen model, follow the general procedure below:

a. **Break a bond** between two atoms and take note of the change in energy in the model. Record your observations and data in the chart below.

b. **Form a bond** between two atoms and take note of the change in energy in the model. Record your observations and data in the chart below.

Data	Observations

ANALYSIS AND INTERPRETATION

6. What patterns do you observe in the data you collected above?

7. How is energy transformation in this system?

BUILDING UNDERSTANDINGS DISCUSSION

8. Draw a picture showing the stocks and flows of energy during each process.

	Bond Breaking	Bond Making
Before		
After		

9. When is energy going into the system? What evidence from your investigation supports this?

10. When is energy going out of the system? What evidence from your investigation supports this?

11. How have we decided to talk about the types and flows of energy in the system?

MODEL REVISION

12. We have been physically manipulating our models and putting energy into the system to rearrange particles. However, we know that realistically atoms are too small to be manipulated and rearranged in this way and therefore our current model is limited in its accuracy. **How do atoms normally move and interact? How can we modify our model to represent that movement more accurately?**

13. How can we more precisely record and keep track of changes in particle speed beyond our own observations?

INVESTIGATION PART 2

14. Record your observations and data below:

	Bond Breaking	Bond Making
Data & Observations		

TRACKING ENERGY CHANGES

15. On the grid below, make a rudimentary graph to show the **changes in kinetic energy (speed of particles) over time** we saw when **bonds were breaking**.

16. On the grid below, make a rudimentary graph to show the **changes in kinetic energy (speed of particles) over time** we saw when **bonds were forming**.

CONSENSUS DISCUSSION

17. Summarize what we have figured out today in our incremental modeling tracker. **Consider how energy and speed show up in the system of rearranging atoms.**

NEXT STEPS

18. How might our model imprecisely measure what would happen in an actual system of rearranging atoms? Consider how our graphs from questions 15 and 16 could be more accurate.

19. How might our model misrepresent what would happen in an actual system of rearranging atoms? Consider variables that might exist in our model that don't exist with rearranging real atoms.

20. What different type of model could we use to more accurately measure and represent an actual system of rearranging atoms?

21. Imagine you hired a programmer to design a computer simulation for you to replace each of the materials you used with a virtual version of them, so that they interacted in a manner similar to those in the marble experiments, but more like real atoms.

Come up with some design specifications for this virtual model.

Design specifications for your system	
What materials would you need to represent from our marble experiments in order to simulate both bond making and bond breaking?	
Because this is a simulation, you could also program the software to remove or amplify the effect of some interactions. **What kind of interactions between parts of the system would you want to be able to amplify or remove?**	
It was sometimes difficult to keep track of all the pieces of our marble track system. **What kind of measurement tools would you want to build into the system?**	

Lesson 5: Options for Modeling Bond Breaking / Making

Physical Representation	How is this connection similar to the connection between particles?	How is this connection different from the connection between particles?	How does this connection represent kinetic energy when particles rearrange?
Ball and Stick Models with Plastic Connectors 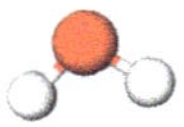 Image: MolView.org	The prongs on the plastic atoms represent electrons being shared or transferred.	Real atoms don't connect with plastic connectors.	We have to move them with our hands to represent kinetic energy.
Bar Magnets  Image by: Jalaldn / CC BY-SA	The north and south poles of the magnet are analogous to the positive and negative charges in atoms.	In real atoms, the positive charge is in the center (protons in the nucleus) surrounded by negative electrons.	Speed of magnet poles attracting each other represents an increase in kinetic energy.
Magnetic Marbles  Image by: Melissa Campanella for iHub	The magnetic field makes the marbles attract like the electrostatic field makes atoms attract.	They are round, in a shape more like real atoms.	Speed of marbles when attracted to magnets represents an increase in kinetic energy.
Paper Cut Outs of Atoms 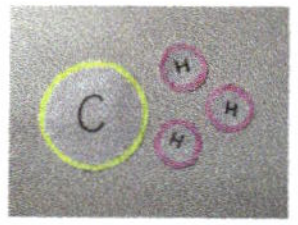 Image by: Michelle Zhang for iHub	We can represent different types of atoms and how they rearrange in chemical reactions.	We have to move them with our hands to make them connect.	We have to move them with our hands to represent kinetic energy.

NAME: ____________________

PERIOD: __________ DATE: __________

Lesson 6: Why might the breaking and making of bonds lead to temperature changes?

EXPLORING THE SIMULATION

1. Watch the screencast of the simulation (sim) and jot your noticings and wonderings below.

Notice	Wonder

2. Compare the sim to your magnetic marble model from the previous lesson. Which parts or actions/movements within the sim correspond to which parts or actions/movements with the model?

3. Now think about how the sim represents bond breaking and bond making. Draw a picture or describe the parts and actions/movements within the simulation that show:

A. When bonds are broken.	
B. When bonds are made.	
C. The reactants and products.	
D. Data on how kinetic energy of particles is changing.	
E. The energy associated with electrostatic fields between particles.	

EVALUATING THE SIMULATION

4. Open the sim at this **link** (tinyurl.com/BondsandEnergyNetlogo) and play with the controls.

5. Go back to your list of design specifications from Lesson 5 and **put a checkmark** next to each thing that appears to be part of this simulation that was on your list.

6. How well does the simulation match your specifications? What does the simulation allow us to do that the marble model of bond making and breaking doesn't?

7. What might be some limitations of this simulation as a model of what is going on when atoms rearrange in a chemical reaction?

CONSENSUS BUILDING DISCUSSION

8. Decide with the class what questions we want to answer using the simulation. List the final questions the class agrees on below.

- ____
  ____
  ____?
- ____
  ____
  ____?
- ____
  ____
  ____?

9. How might answering these questions help us figure out why hydrogen fuel has such a high energy output compared to gasoline? Are there any other questions on our DQB this might help answer?

PLANNING THE INVESTIGATION

10. Break into groups. Then decide with your group which of the above questions you'd most like to answer in your investigation. Write this question below or star it * above.

Note: Work with the class to make sure that at least one, and ideally more than one team is working to answer each question.

PLANNING THE INVESTIGATION, CONT'D

11. Use the questions below to help you plan an investigation to answer your group's focal question. As you discuss, **keep a record of your answers and** make sure **each team member's ideas** are represented.

- Brainstorm with your group: What might be some different ways you could use the simulation to answer this question?
- Evaluate the ideas you've generated. Which approach will best answer your question and why?
- List the variables you need to pay attention to in your investigation in order to answer your question.
- Which of these variables are dependent or independent? How do you know?
- Which variables need to be controlled (kept the same) and why? How will you set up the sim to keep those variables controlled?
- Which variables will you need to manipulate or change to answer your question and why? How will you adjust the sim to make those variables change?
- How will you keep track of your observations and measurements to produce data that will answer your question?

CONDUCTING THE INVESTIGATION

12. Use your plan to conduct your investigation and gather data to answer your group's question. Be ready to share out an answer to your question and supporting data.

GOING PUBLIC

13. Agree as a class on a plan that will ensure 1) every team is able to publicly report out an answer to their focal question and share data they gathered that informed their thinking, AND 2) all teams are able to create a record that includes an answer and supporting data for each question the class decided to investigate. You can use the space on the next page to capture notes.

Class Consensus Questions	Answer to Questions & Supporting Data
Q1	
Q2	
Q3	

MAKING SENSE OF OUR INVESTIGATIONS

14. In the reactions we've seen, bonds are both broken and formed. Keeping this in mind, why do you think burning (gasoline, ethanol, or hydrogen fuel) resulted in energy being released?

15. If all chemical reactions require breaking some bonds and forming new bonds, why do some reactions result in an increase in energy to the surroundings (hotter) while others result in a decrease in energy to the surroundings (cooler)? Why doesn't the energy used to break bonds just balance out with the energy released when bonds are formed (resulting in a constant temperature)?

16. How might we investigate this question further with our computer model?

17. Summarize what you are thinking now in your **Incremental Modeling Tracker (IMT)**.

Reference:

Sengupta, P. and Wilensky, U. (2005). NetLogo Electrostatics model. http://ccl.northwestern.edu/netlogo/models/Electrostatics. Center for Connected Learning and Computer-Based Modeling, Northwestern University, Evanston, IL. - Wilensky, U. (1999). NetLogo. http://ccl.northwestern.edu/netlogo/. Center for Connected Learning and Computer-Based Modeling, Northwestern University, Evanston, IL.

Investigations are from the Northwestern University Storyline "Why do some things get colder or hotter when they react?"

NAME: ______________________________

PERIOD: ____________ DATE: ____________

Lesson 7: Why doesn't the energy that goes IN to break bonds and that comes OUT when bonds are made just balance out?

RECAP

1. What did we figure out from our investigations using the NetLogo simulation **last class?**

- How does bond breaking affect the overall energy in a reaction system?
- How does bond making affect the overall energy in a reaction system?
- What causes overall changes in energy to be big or small?

PLANNING OUR INVESTIGATION

2. Last time we considered how we might use our simulation to answer the question "Why doesn't the energy that goes IN to break bonds and that comes OUT when bonds are formed just balance out?"

3. What ideas did you have? Add to or clarify them with the class.

INVESTIGATION

4. Let's try to figure it out!

- Can you get the **overall energy in the system to go UP** by both breaking and forming bonds? Under what **conditions?**

- Can you get the **overall energy in the system to go DOWN** by both breaking and forming bonds? Under what **conditions?**

- Can you get the **overall energy in the system to stay CONSTANT** by both breaking and forming bonds? Under what **conditions?**

Question	Investigators	Answer to questions & Supporting data
Question 1		
Question 2		
Question 3		

MAKING SENSE

5. Scenario 1: The magnets are equal strength. For example:

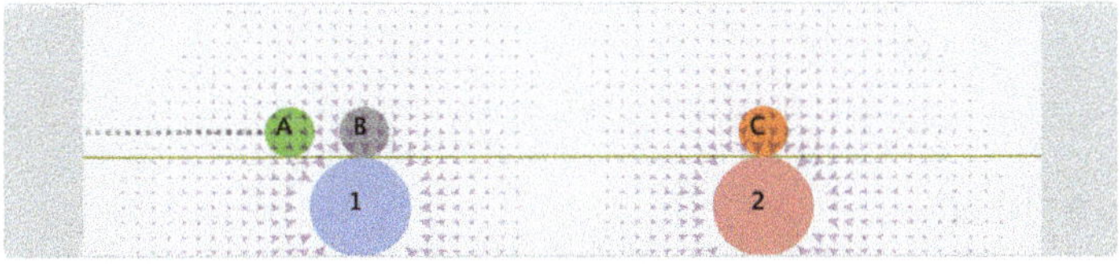

Predict: How would the overall kinetic energy in the system change in Scenario 1? Why?

____

____

6. Scenario 2: The strength of magnet 1 is **weaker** than the strength of magnet 2.

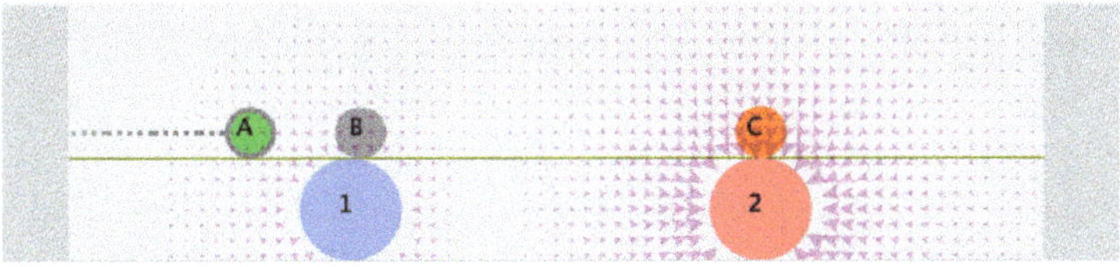

Predict: How would the overall kinetic energy in the system change in Scenario 2? Why?

____

____

7. Scenario 3: The strength of magnet 1 is **stronger** than the strength of magnet 2.

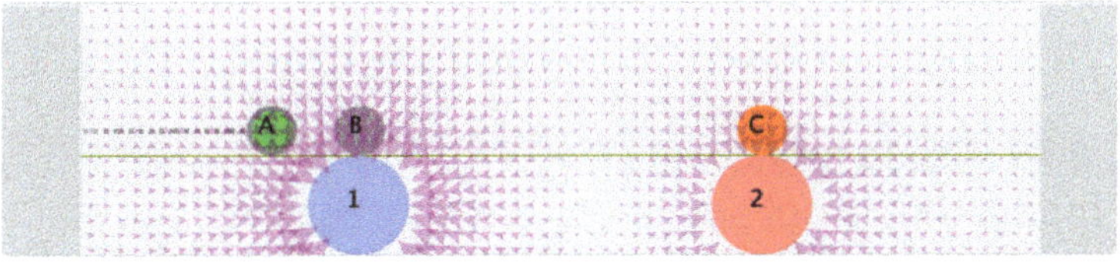

Predict: How would the overall energy in the system in Scenario 3? Why?

____

____

8. Are there **any other questions from our DQB we answered using our simulation today?**

SMALL GROUP MODELING / EXPLANATIONS

9. Use what you figured out in the last 4 investigations to **explain and/or model on a separate sheet why the barium hydroxide and ammonium chloride reaction resulted in a drop in temperature while burning fuel (either gasoline or ethanol) resulted in an increase in temperature**.

Be sure to include:

- A claim about which scenario/set of conditions in the simulation you think corresponds to what is happening at the atomic level in each reaction.
- An explanation for why those atomic scale conditions would result in the macro-scale energy change we observed in the reaction.
- Key words or phrases that will help others understand your group's thinking.
- Ideas of everyone in the group.

MODEL GALLERY WALK

10. Post your model/explanation, then walk around the room to look at the other group's work. Use sticky notes to leave feedback.

11. After returning to your table, read through your group's feedback and reflect on what you saw. What do you need to add or change about your model?

WHOLE CLASS CONSENSUS MODEL / EXPLANATION

12. As a class, try to make a ***general*** model that is not specific to any one reaction but still explains why some chemical reactions result in a drop in temperature while others result in an increase in temperature increase.

You might want to have one model for chemical reactions that result in a drop in temperature and a second model that explains chemical reactions that result in an increase in temperature.

Make sure your model includes a way of seeing a difference between the energy before the chemical reaction and the energy after the chemical reaction.

CONSENSUS MODEL / EXPLANATION

Why do some chemical reactions result in a drop in temperature while others result in an increase in temperature increase?

13. During the discussion, answer these questions:

a. What are all the different ways we've been talking about [energy flow, bonds, temperature]?

b. What have we decided to call reactions that result in a decrease in temperature?

____

c. What have we decided to call reactions that result in an increase in temperature?

____

d. What have we decided to call the amount of energy required to break a bond?

____

e. What determines if a reaction results in an increase or decrease in energy to surroundings?

f. Consider our current definition of a chemical reaction. Is it possible to have a chemical reaction that is not exothermic or endothermic? If so, how?

g. Do we have any way of predicting how much energy will be released when a bond is formed?

h. After seeing your classmates presentations, does our class consensus model actually explain all of the phenomena that we saw in this unit? Why or why not?

i. What happens to **matter** in these reactions? Does it change with the energy transformations?

NEXT STEPS

14. Take a moment to make some notes in the Incremental Modeling Tracker. **What do we need to figure out next? How should we investigate this?**

NAME: ____________________

PERIOD: __________ DATE: __________

Lesson 8: What makes some bonds stronger and other bonds weaker? Why do atoms bond anyway and what's 'stability'?

REVIEW

1. Describe what we figured out in the last lesson regarding bonds and bond strengths. Then, predict *why different bonds might have different bond strengths*?

DATA ANALYSIS

2. Look at the bond energies chart in the supplemental materials.

Noticings	Wonderings

MODELING BOND LENGTH

3. We've got lots of different models we've been using so far to represent bonds.

Magnetic Marbles Model of Bonding	Netlogo Simulation
	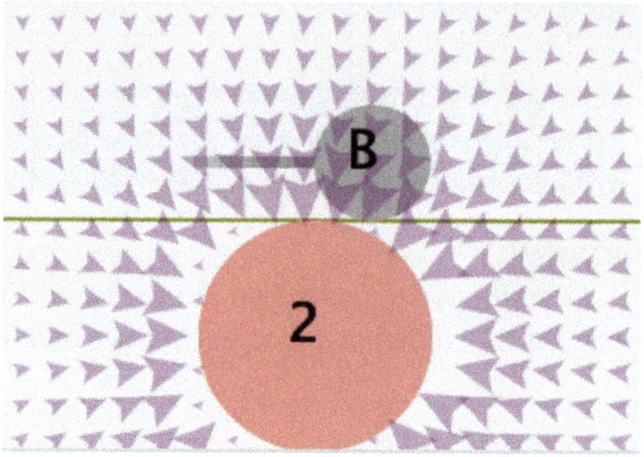

We also used some different models before for modeling water:

Octet Rule Model of Bonding
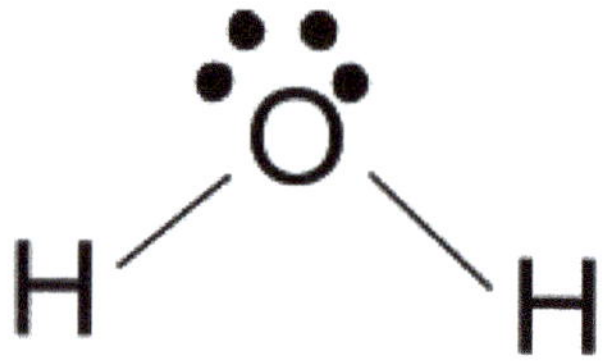

How might we revise our magnetic marble model to be more consistent with what we know about atomic structure?

In this proposed model, what do the different components represent and how they interact within the system?

DEMONSTRATION

An inverter magnet includes a plastic ring that can hold a large magnet in the center and six smaller magnets around the edge. Examine the diagram of the inverter magnet system.

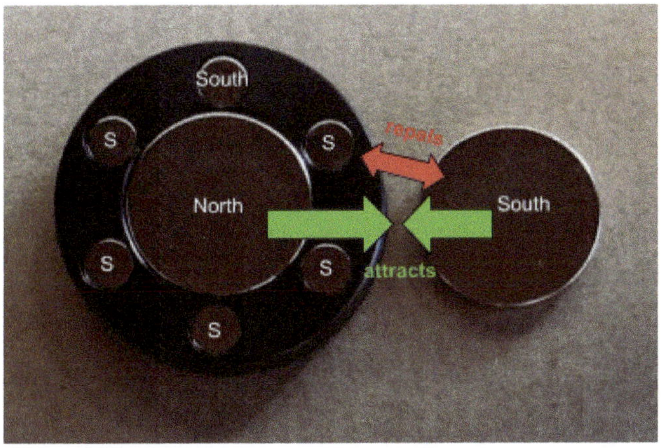

How might this be similar to or different from the structure of an atom?

When is this **system** most **stable**?

4. Watch the demo with the inverter magnet and record your observations and data below.

As you watch, pay particular attention to:

- Patterns: What do you see happening in predictable ways with the inverter magnet?
- Stability & change: What becomes different and what stays the same?
- Energy & matter: How might the energy of the magnetic field be changing as the position of the magnets changes? Why?

Quantitative Data	Qualitative Data

5. What are some **strengths and limitations** of using the inverter magnet to represent bonding and atomic structure?

THOUGHT EXPERIMENTS

Atoms are composed of positively charged nuclei surrounded by a negatively charged cloud of electrons. When bonds form, the nuclei of one atom attract the electrons of another atom, pulling the atoms toward each other.

Thought Experiment 1: Imagine two bonded atoms being pushed even closer together, so much so that their nuclei are almost touching.

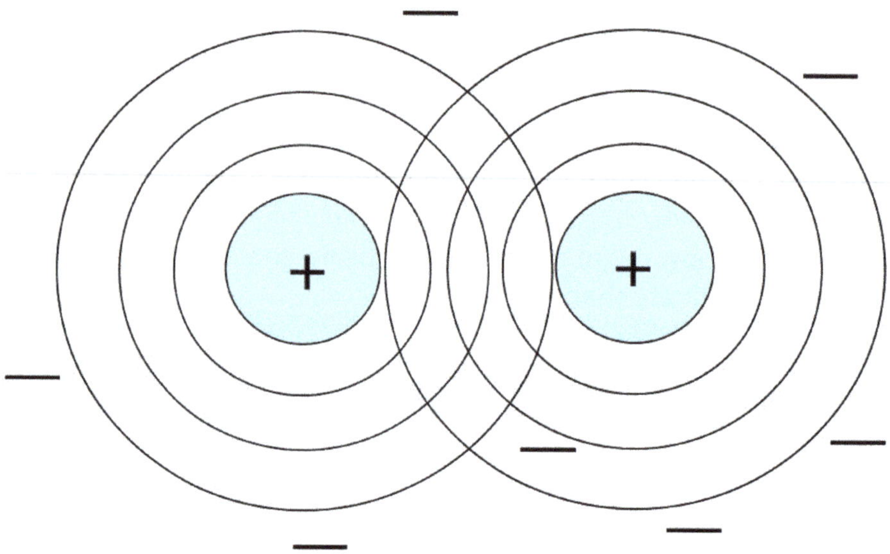

Image: Created by MH for inquiryHub

6. What would happen? Why?

Thought Experiment 2: Imagine we switched out a magnet in the inverter system to a stronger or weaker magnet.

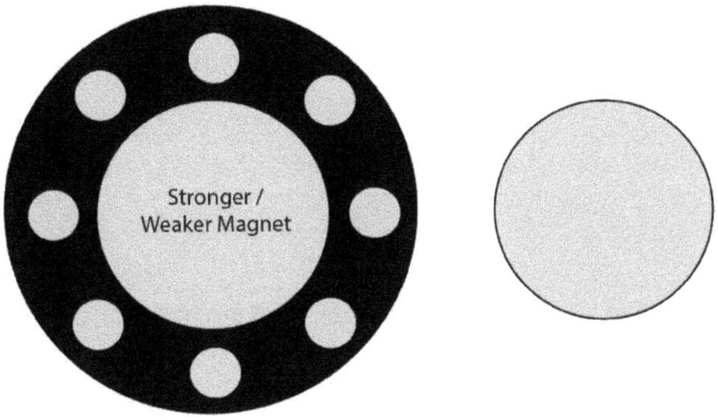

Image: Created by MH for inquiryHub

7. Would the distance between the center and the outer magnets change? If so, how and why?
If stronger:

If weaker:

8. What would happen to the speed of the outer magnet if you let it go to form a "bond"? Why?
If stronger:

If weaker:

9. What would happen to the amount of energy required to break the "bond"? Why?
If stronger:

If weaker:

MAKING SENSE

Could anything we know about atomic structure help us explain **why bonds between some pairs of atoms are stronger and bonds between other pairs of atoms are weaker**? Compare the bond energies chart to the periodic table and electron energy shell diagrams.

10. What patterns do you see in the data presented in the periodic table and energy shell diagrams?

11. In general, how are bond energy and length changing as the bonding atom moves farther down a group in the periodic table?

12. Does the pattern in the data support the conclusion that bond energy is related to bond length? Why or why not? If you notice exceptions, what do we know about atomic structure that could potentially help explain those apparent exceptions?

13. Create a model or write an explanation for why hydrogen's bond with iodine is weaker than hydrogen's bond with chlorine.

14. How do you **predict the bond length and energy would change** if hydrogen formed a bond with Astatine (At)?

15. Draw a picture to show the energy associated with a bond with a long length vs a bond with a short length.

16. Below is a diagram of a bond forming. **What things stay the same** during bond formation? **What things change** in bond formation?

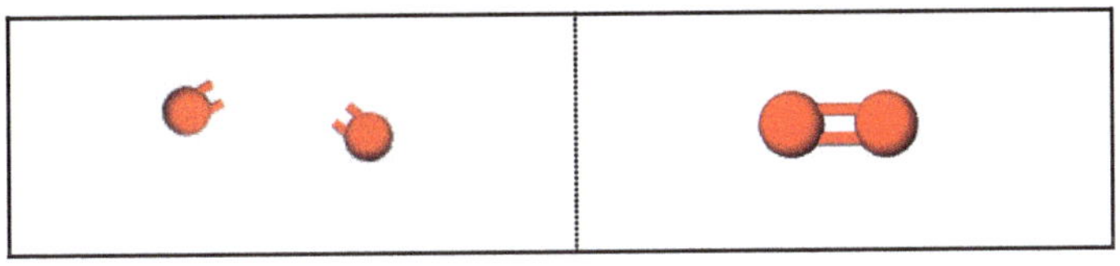

Images from: MolView

17. Before the bond forms, is the system **stable** or **unstable**? Present evidence of what is happening at the atomic scale to support your claim.

18. **After** the bond forms, is the system **stable** or **unstable?** Present evidence of what is happening at the atomic scale to support your claim.

19. Think back to your drawing in question 15. Which bond is **more stable**? Why?

BUILDING UNDERSTANDINGS

20. In a chemical reaction, **how can you predict what will change about the lengths and strengths of bonds being formed vs. broken**? (Hint: the word "stable" might help with your response. You can draw to answer this question.)

21. The octet rule says that atoms form bonds to fill their valence shells and achieve stability. **How can what we figured out today help us predict stability?**

22. Summarize what we have figured out today on your **Incremental Modeling Tracker (IMT)**.

23. Do any of our other investigations help us think about bond length/strength in any **different ways** than the inverter magnets?

24. What do we still need to figure out to explain why burning hydrogen releases more energy than burning octane? How might we go about doing that?

Lesson 8: What makes some bonds stronger and other bonds weaker? Why do atoms bond anyway and what's 'stability'?

Unit Question: Why do cars run on gasoline and nct rocket fuel?

Lesson 8: Periodic Table Showing Electron Shells and Bond Energy Chart

Periodic Table of Elements Showing Electron Shells

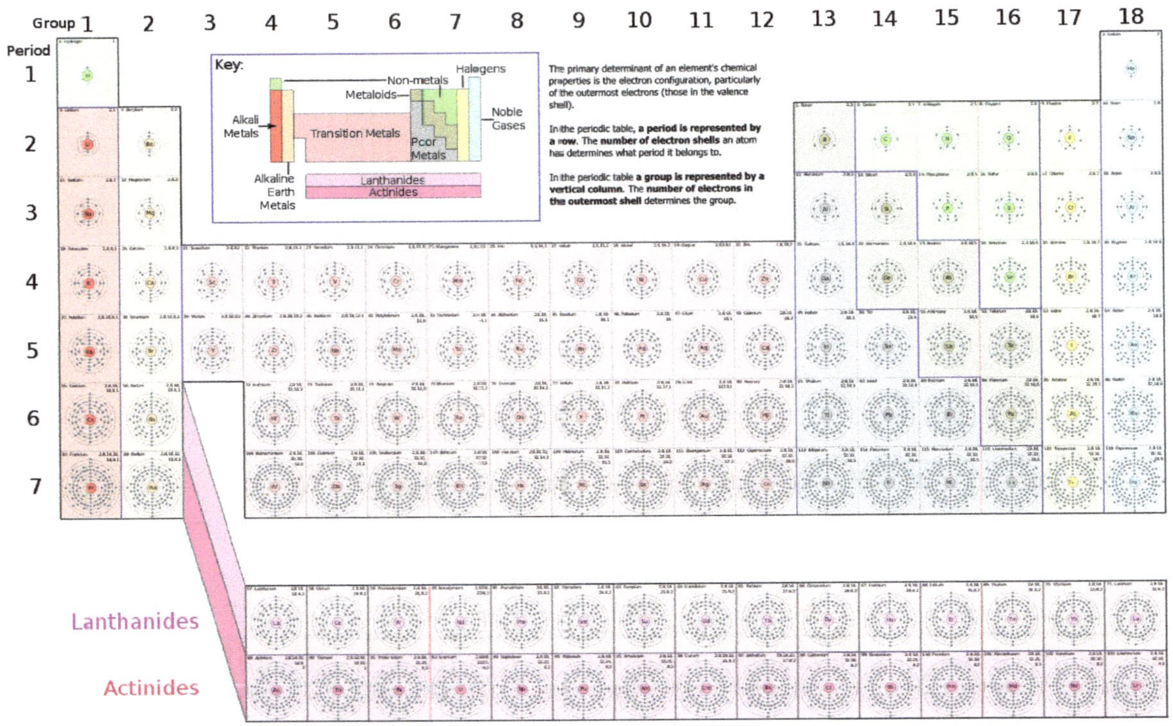

	Bond	Energy (kJ/mol)	Length (pm)
9 F			
17 Cl	H — H	432	75
	H — F	565	92
35 Br	H — Cl	427	127
	H — Br	363	141
53 I	H — I	295	161

Image Source: Greg Robson with Inkscape / CC BY-SA 2.0 UK

These materials were developed by the inquiryHub RPP, including educators from CO and IL, Northwestern University and the University of Colorado Boulder, with support from the Hewlett Foundation, the Spencer Foundation, the National Science Foundation, and Denver Public Schools.

Bond Energy Chart

Bond Lengths and (Average) Bond Dissociation Energies for some Common Covalent Bonds

Bond	Energy (kJ/mol)	Length (pm)	Bond	Energy (kJ/mol)	Length (pm)	Bond	Energy (kJ/mol)	Length (pm)
H — H	432	75	N — H	391	101	F — F	154	142
H — F	565	92	N — F	272	136	F — Cl	253	165
H — Cl	427	127	N — Cl	200	175	F — Br	237	176
H — Br	363	141	N — Br	243	189	F — I	273	191
H — I	295	161	N — N	160	145	Cl — Cl	239	199
C — H	413	109	N = N	418	125	Cl — Br	218	214
C — F	485	135	N ≡ N	941	110	Cl — I	208	232
C — Cl	339	177	O — H	467	96	Br — Br	193	228
C — Br	276	194	O — F	190	142	Br — I	175	247
C — I	240	214	O — Cl	203	172	I — I	149	267
C — C	347	154	O — I	234	206	S — H	363	134
C = C	614	134	O — O	146	148	S — F	327	156
C ≡ C	839	120	O = O	495	121	S — Cl	253	207
C — N	305	143	C — O	358	143	S — Br	218	216
C = N	615	138	C = O	745*	120	S — S	226	205
C ≡ N	891	116	*C = O bond energy of CO_2 is 799 kJ/mol			Adapted from: Chegg.com		

NAME: ____________________

PERIOD: __________ DATE: __________

Lesson 9: Can we explain now why hydrogen has such a high energy output per gram compared to other fuels?

LOOKING BACK

1. Review your **incremental modeling tracker (IMT)**. What are all of the **pieces** that we've figured out so far that would help us finally explain how it is that hydrogen produces so much more energy per gram than octane?

CLASS RECORD

As the group shares, use the space below to **capture ideas** that will be important **to include in our model/explanation.**

BUILDING A PHYSICAL MODEL

Use the supplemental materials document as a resource as you build models representing the combustion of each fuel, Hydrogen, and Octane. You are also welcome to use the atom cutouts at the end of this document in your model if this makes sense.

Decide on a way to document your model. (Photos? Videos? Sketches?) Ensure your representation includes the criteria we decided on together as a class.

Document your observations and thoughts about energy and matter as you model both reactions in the graphic organizer below:

	Hydrogen combustion	Octane combustion
What bonds are being **broken**? (Ex. O-H, C-C) **How many of each**?		
What bonds are being **made**? (Ex. O-H, C-C) **How many of each**?		
Jot any **noticings** you have about the amounts of **matter and energy** involved in each reaction.		
Jot any **wonderings** you have about the amounts of **matter and energy** involved in each reaction.		

BUILDING A COMPUTATIONAL MODEL

Build a **computational model** - a calculation, series of calculations, or equation(s) - to accompany your model from the previous step. Consider the following graphic that defines the system and surroundings of a chemical reaction, then outline your computational model in the graphic organizer.

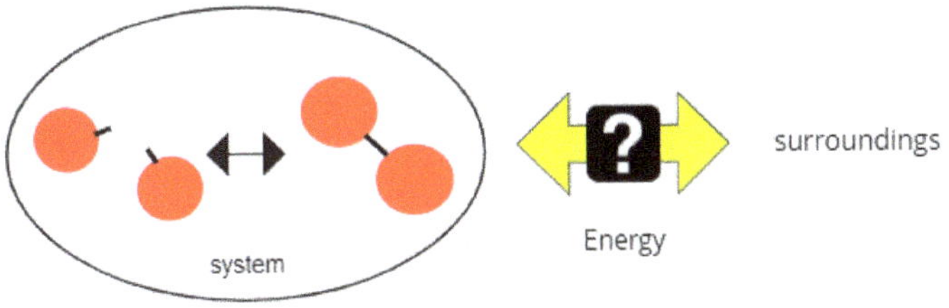

Image: Michelle Zhang for iHub

<table>
<tr><th>Breaking bonds</th><th>Making bonds</th></tr>
<tr><td>If breaking bonds requires energy IN, or takes energy, what mathematical function (+, −, ×, ÷) should we use to represent this process in a computational model? Why?</td><td>If making bonds results in energy OUT, or releases energy, what mathematical function (+, −, ×, ÷) should we use to represent this process in a computational model? Why?</td></tr>
<tr><td>How could we keep track of all the energy involved in breaking bonds?</td><td>How could we keep track of all the energy involved in making bonds?</td></tr>
<tr><td colspan="2">Note: Remember, you already kept track of each bond that was broken or made in your physical model. We also know the average bond dissociation energies of various bonds from the chart on the supplemental resources document.</td></tr>
<tr><td>If there is more energy involved in breaking bonds than making bonds, what do we expect to see in the temperature of the reaction? Why? (Breaking energy > making energy)</td><td>If there is more energy involved in making bonds than breaking bonds, what do we expect to see in the temperature of the reaction? Why? (Breaking energy < making energy)</td></tr>
</table>

USING A COMPUTATIONAL MODEL

2. Predict: Should we expect to see a **positive or negative value** at the end of our calculations for **both combustion reactions**? Why?

3. Predict: Which reaction do we expect will result in the largest energy changes? Why?

Your computational model should 1) show energy flows and changes as atoms rearrange to form new bonds in each combustion reaction and 2) help us explain the difference in energy output when burning Hydrogen fuel vs. Octane fuel.

Try out your computations using the graphic organizer below and the supplemental materials document as a resource. Then, represent your computations on chart paper or a virtual platform where it can be publicly shared.

Breaking bonds	Making bonds

MAKING SENSE

4. How would you **summarize the puzzle we ran into** when using our computational model?

5. What do you think might be going on here?

Lesson 9: Part 2 - Can we explain now why hydrogen has such a high energy output per gram compared to other fuels?

SEEDING A NEW PROBLEM

Reflect on the two charts that you were shown:

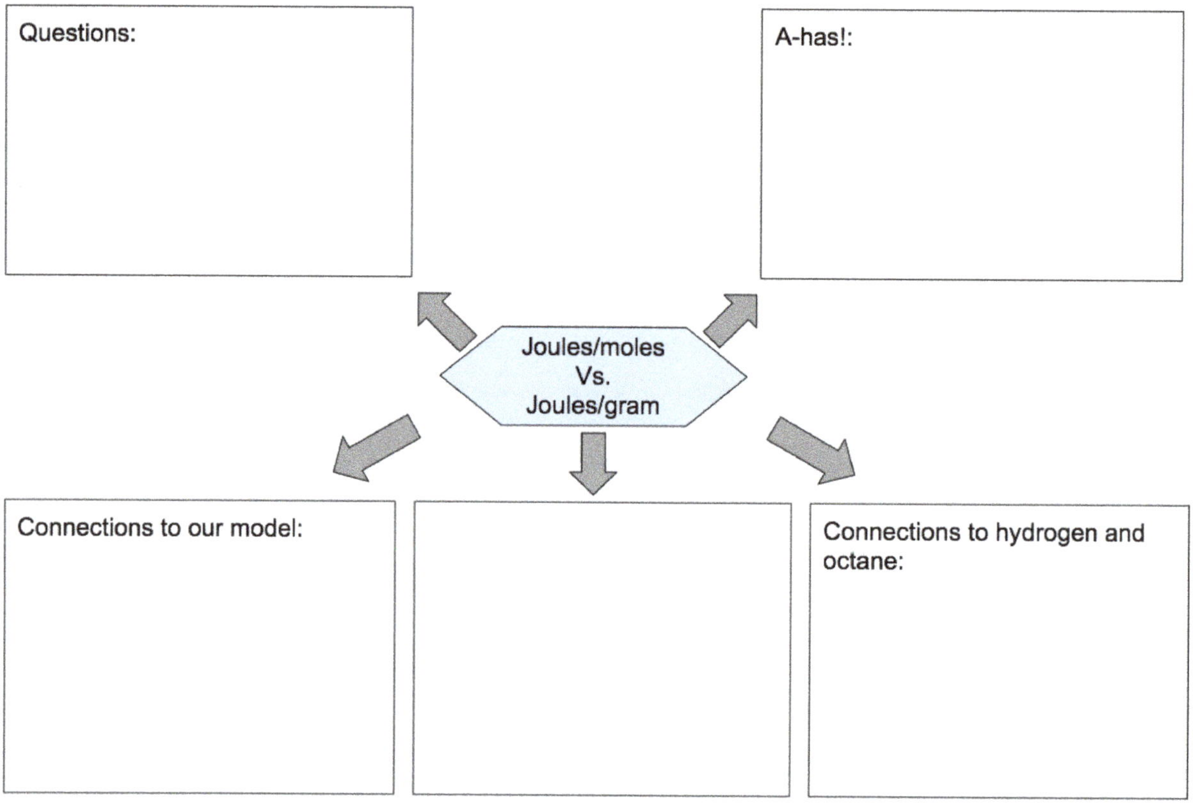

NEXT STEPS

6. What questions do we still have after today's lesson?

7. What should we investigate next to explain our unanswered questions? **How** should we investigate it?

8. Take a moment to update your **incremental modeling tracker** (IMT).

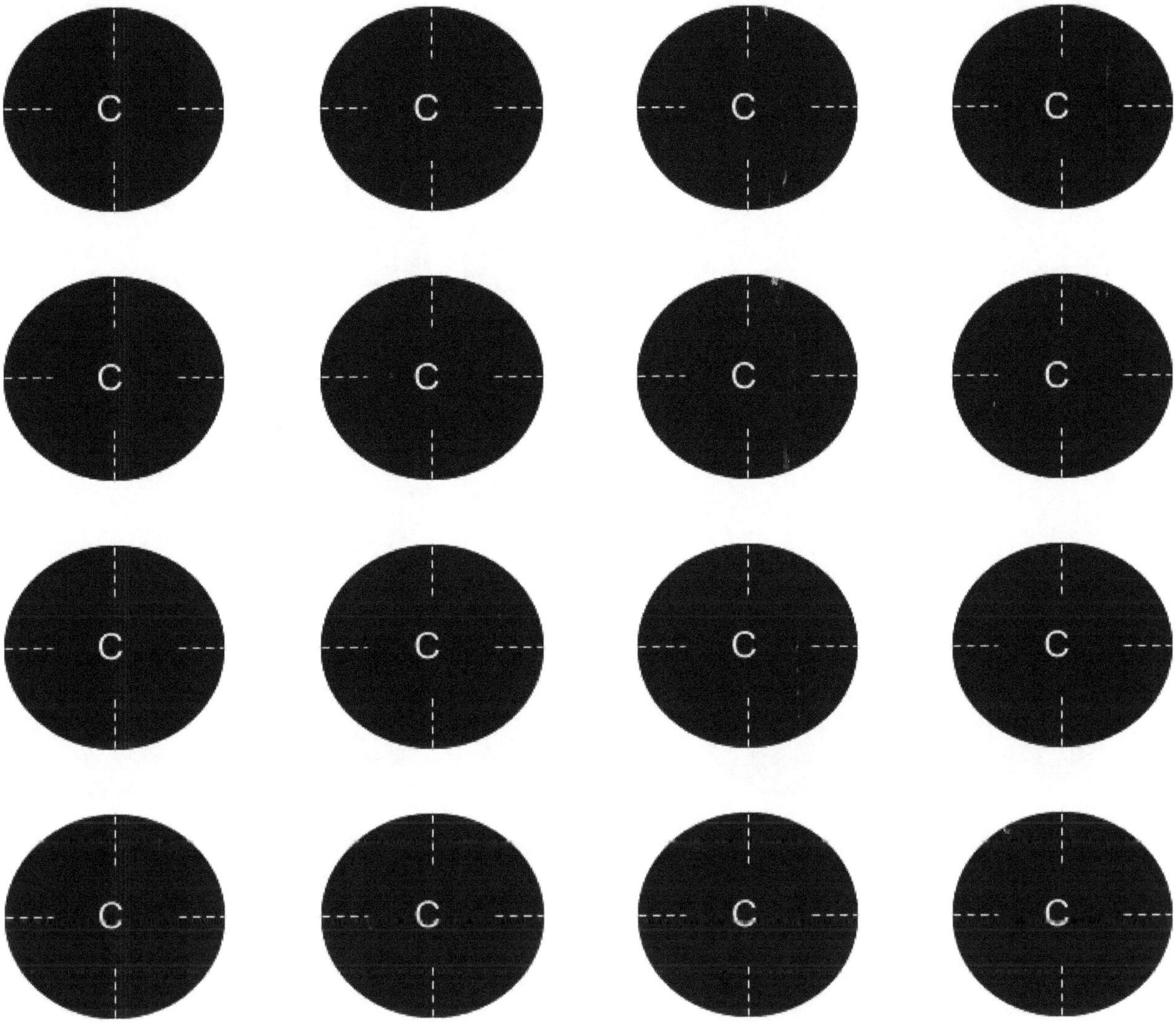
C
C
C
C
C
C
C
C
C
C
C
C
C
C
C
C

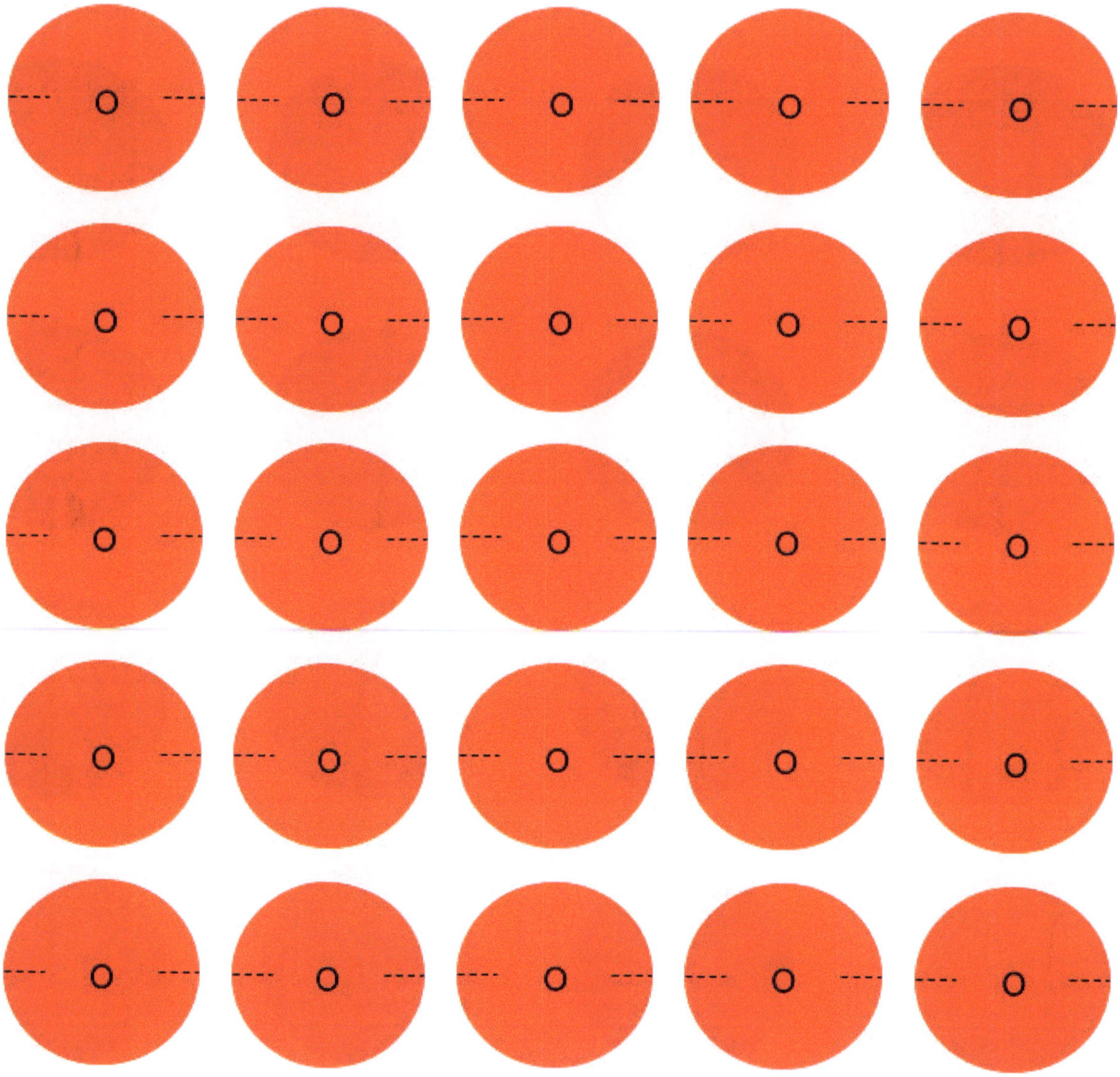
o o o o o
o o o o o
o o o o o
o o o o o
o o o o o

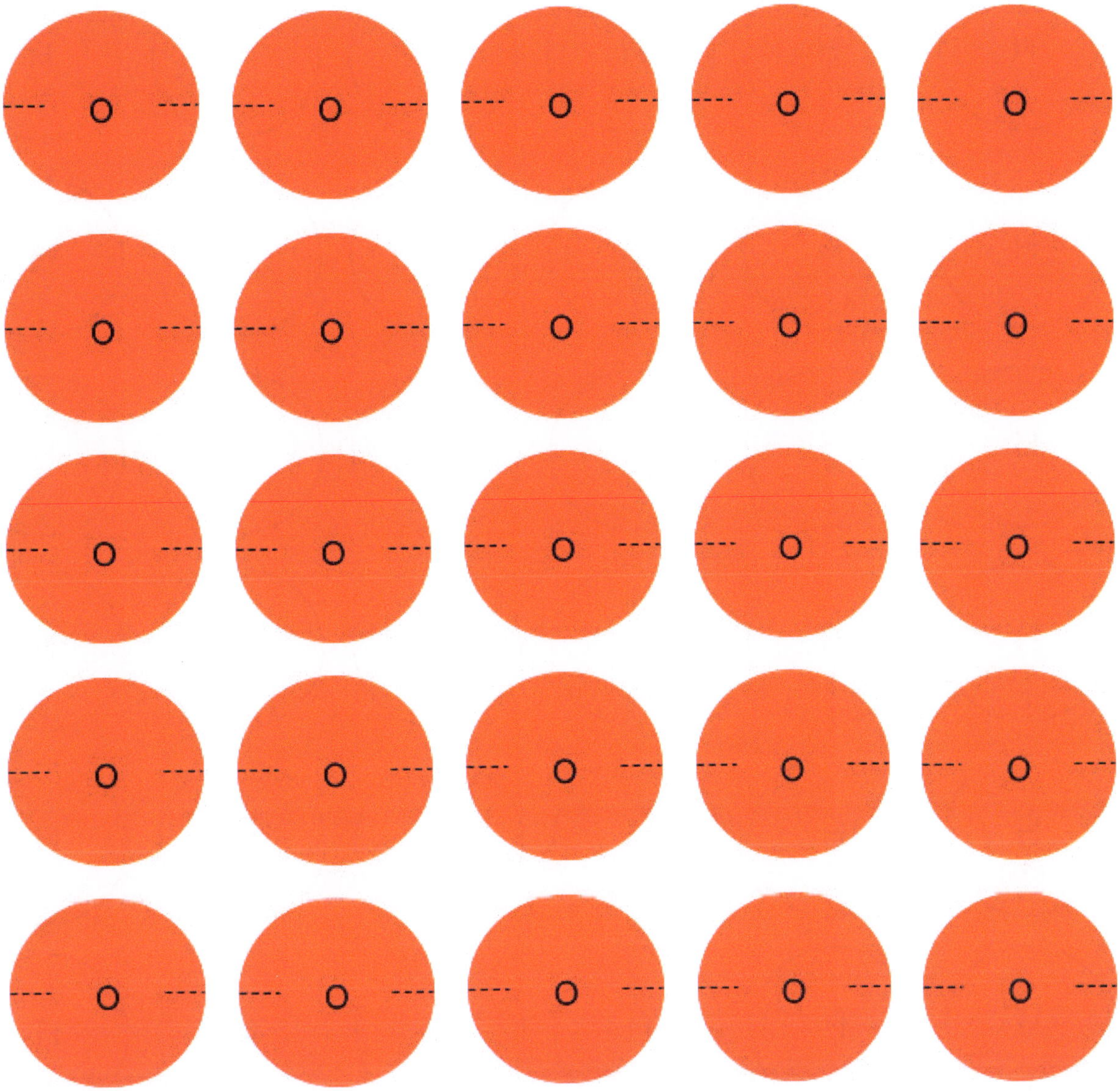
o o o o o
o o o o o
o o o o o
o o o o o
o o o o o

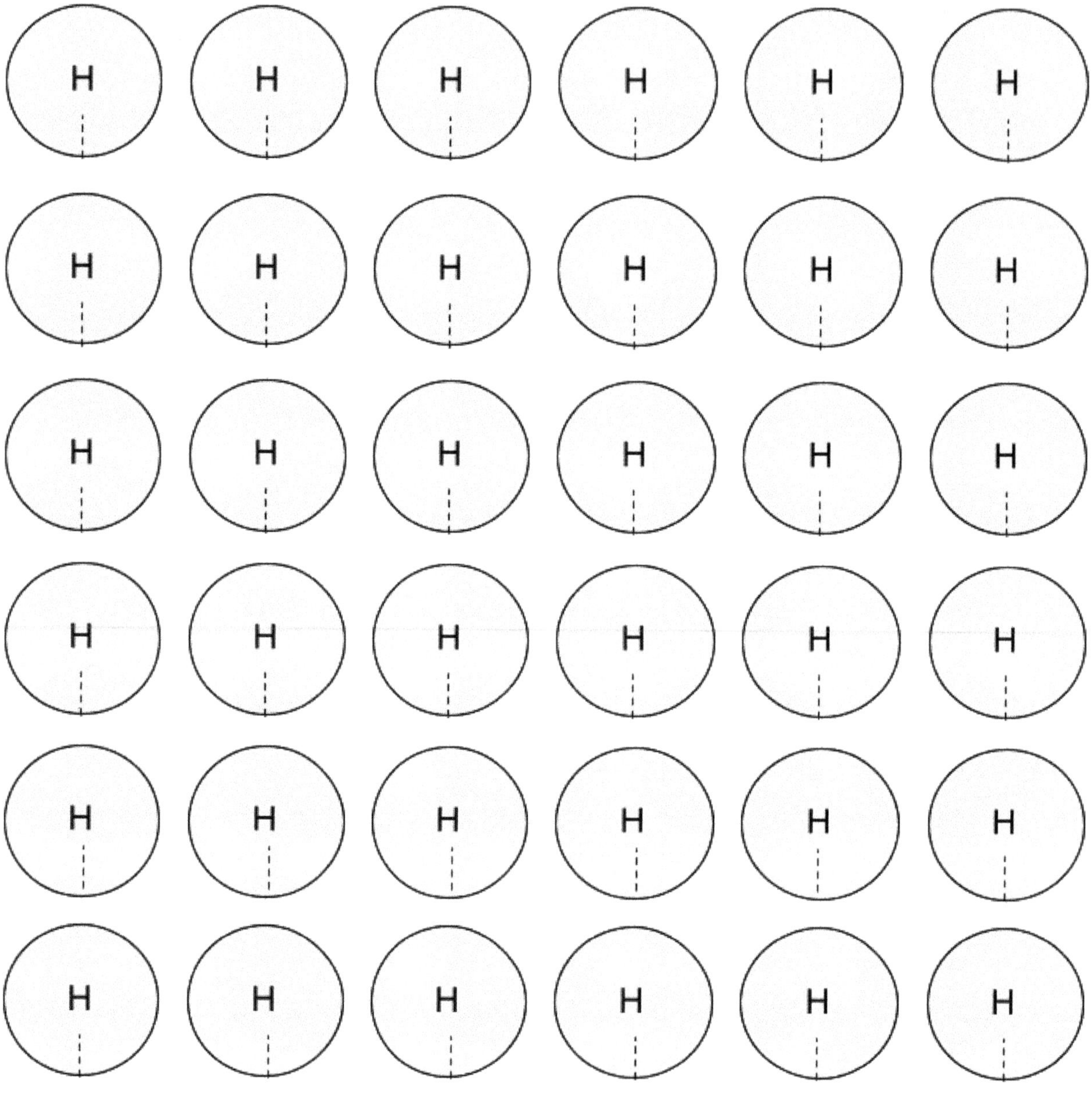
H H H H H H
H H H H H H
H H H H H H
H H H H H H
H H H H H H
H H H H H H

Lesson 9: Can we explain why hydrogen has such a high energy output per gram compared to other fuels?

NAME: ______________________

Unit Question: Why do cars run on gasoline and not rocket fuel?

PERIOD: ____________ DATE: ____________

Lesson 9: Reaction Equations & Molecule's Structural Formulas & Data Charts

Fuel	Combustion Reaction	Energy (kJ/g)	CO_2/gal (lbs)
Gasoline	$2C_8H_{18} + 25O_2 \longrightarrow 16CO_2 + 18H_2O$	46.5	20
Hydrogen	$2H_2 + O_2 \longrightarrow 2H_2O$	141.9	0

Bond Lengths and (Average) Bond Dissociation Energies for Common Covalent Bonds in Combustion Reactions

Bond	Energy (kJ/mol)	Length (pm)	Bond	Energy (kJ/mol)	Length (pm)	Bond	Energy (kJ/mol)	Length (pm)
H — H	432	75	O — O	146	148	C — C	347	154
C — H	413	109	O = O	495	121	C = C	614	134
O — H	467	96	C — O	358	143	C ≡ C	839	120
C = O	745*	120	***C = O bond energy of CO_2 is 799 kJ/mol**			Adapted from: Chegg.com		

Molecule	Formula	Structure
Hydrogen	H_2	
Oxygen	O_2	
Water	H_2O	
Carbon Dioxide	CO_2	
Octane	C_8H_{18}	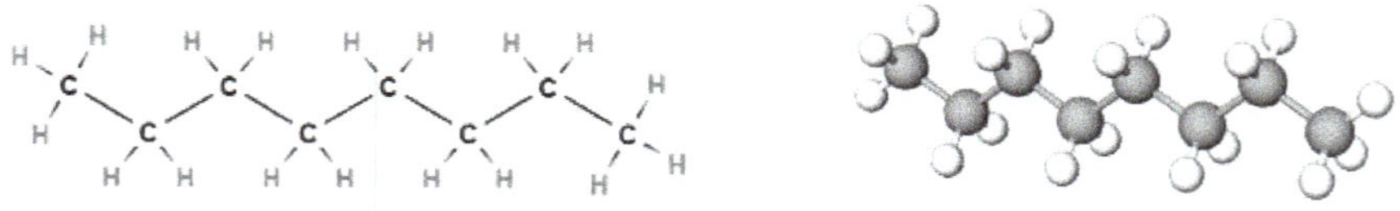

Images from: MolView.org

Bond Lengths and (Average) Bond Dissociation Energies for some Common Covalent Bonds

Bond	Energy (kJ/mol)	Length (pm)	Bond	Energy (kJ/mol)	Length (pm)	Bond	Energy (kJ/mol)	Length (pm)
H — H	432	75	N — H	391	101	F — F	154	142
H — F	565	92	N — F	272	136	F — Cl	253	165
H — Cl	427	127	N — Cl	200	175	F — Br	237	176
H — Br	363	141	N — Br	243	189	F — I	273	191
H — I	295	161	N — N	160	145	Cl — Cl	239	199
C — H	413	109	N = N	418	125	Cl — Br	218	214
C — F	485	135	N ≡ N	941	110	Cl — I	208	232
C — Cl	339	177	O — H	467	96	Br — Br	193	228
C — Br	276	194	O — F	190	142	Br — I	175	247
C — I	240	214	O — Cl	203	172	I — I	149	267
C — C	347	154	O — I	234	206	S — H	363	134
C = C	614	134	O — O	146	148	S — F	327	156
C ≡ C	839	120	O = O	495	121	S — Cl	253	207
C — N	305	143	C — O	358	143	S — Br	218	216
C = N	615	138	C = O	745*	120	S — S	226	205
C ≡ N	891	116	*C = O bond energy of CO_2 is 799 kJ/mol			Adapted from: Chegg.com		

Lesson 10: What are the similarities and differences between one mole and one gram of hydrogen and gasoline?

Unit Question: Why do cars run on gasoline and not rocket fuel?

NAME: ______________________

PERIOD: __________ DATE: __________

Lesson 10: What are the similarities and differences between one mole and one gram of hydrogen and gasoline?

REVIEW

1. Draw a model that represents what we learned last class about the differences between hydrogen and octane. Be sure to include energy outputs in your models.

2. Looking at the models you just drew, what question is important for us to answer today?

TODAY'S QUESTION / INVESTIGATION

3. Why is answering today's question important and how will it help us determine why we can't use rocket fuel in cars?

4. Can you think of any investigations we could do in order to figure out the difference between the two different energy output units?

INVESTIGATING THE DIFFERENCES BETWEEN MOLES AND GRAMS WITH SUGAR AND SALT

5. What are your initial observations about a mole of salt compared to a mole of sugar?

6. Find the mass of each of these substances and record them below.

Mass of 1 mole of salt	
Mass of 1 mole of sugar	

7. What initial ideas do you have about the relationship between a gram and the mole after looking at these two substances?

8. Research the mass of a mole of our fuels and record those masses below.

Mass of 1 mole of hydrogen gas	
Mass of 1 mole of octane	

9. Look at the molecular formulas, the ball and stick models, and number of particles in a mole for sugar and salt. What do you notice about them?

10. Using our observations, write a definition for a mole of a substance.

11. What can you infer about the mass of a mole of a substance?

INVESTIGATING THE DIFFERENCES BETWEEN MOLES AND GRAMS WITH HYDROGEN AND OCTANE

12. Now compare the molecular formulas, ball and stick models, and number of particles in a mole for hydrogen and octane. What do you notice about those two substances?

13. Does what you noticed in the previous step support the definition we wrote for a mole? Why or why not?

14. Use the periodic table to support why the mass of one mole of hydrogen is about 2 grams. Can you use the periodic table to show what the mass of one mole of octane is? Does this match the research we did in step 8?

15. Revisit your definition of a mole from step 10. Is there anything you want to revise? Create a final definition for a mole and include its relationship to mass.

16. Record our class definition of a mole below.

MODELING A GRAM AND A MOLE

17. Now that we have a definition of a mole and how it relates to mass, draw a model to show the difference between a gram of salt and a gram of sugar.

18. Draw a model of a representation of a mole of salt and a mole of sugar.

19. When looking at our model of a mole of each substance, what are some of the differences we notice?

NEXT STEPS

20. How can we use what we learned today to make progress on our unit question? Where do we need to go from here?

ADDITIONAL PRACTICE WITH THE MOLE

21. Which of the following statements is true about mole?

a. It is a quantity having mass equal to its atomic/ molecular mass.
b. It is equal to 6.02×10^{23}.
c. It is the chemist's counting unit.
d. All of the above.

22. Since ^{12}C's molar mass is 12 grams, 48 grams of ^{12}C atoms would be equal to _____ moles.

23. CH_4 (methane) has a molar mass of 16. What is the molar mass of each element in the molecule?

24. Water, H_2O is made up of two elements:

a. What are the two elements?

b. What is the mass of one mole of each element?

c. What is the mass of 1 mole of H_2O?

25. Jeniah was calculating the mass of her favorite rocket fuel Hydrogen Peroxide (H_2O_2). She wants to use some of it in a devil's toothpaste explosion, but she needs to do it safely and have accurate measurements. Check her work to ensure she has proper amounts.

Below is her calculation of the molar mass of Hydrogen Peroxide using the periodic table:

Hydrogen -	1 gram per mole x 2 atoms	= 2 grams per mole
Oxygen -	16 gram per mole x 1 atoms	= 16 grams per mole
		Total = 18 grams per mole

Is Jeniah's calculation going to result in the reaction she is planning? How can she improve?

____

____

____

____

____

____

NAME: ______________________

PERIOD: ____________ DATE: ____________

Lesson 11: Why does burning one mole of gasoline result in more energy than burning one mole of hydrogen?

LOOKING BACK

1. Use the ball and stick models you see on the board. If you had a gram of water and a gram of oxygen, which substance would you have more particles of? Why?

2. If you had a mole of each substance which substance would you have more of?

3. How much would a mole of each substance weigh?

4. How many moles of CO_2 would a mole of each fuel produce?

5. How does producing CO_2 impact the effects of the fuel on the environment?

BUILDING A CLASS RECORD

6. Think about all of the concepts we have learned up to this point. If we were going to make a set of 'rules' that we would need to follow in order to model the difference between burning a mole of a fuel vs. burning a gram of a fuel, what would we need to consider?

MODELING BURNING A MOLE VS. A GRAM OF FUEL

7. Create a model depicting burning a mole of hydrogen and a mole of octane. Be sure to note the energy differences and why there are energy differences.

GALLERY WALK

8. What "great ideas" might you want to borrow from other groups?

9. Use your previous two models to create a table model.

NEXT STEPS

10. Now that we know the difference in our energy output units, what questions do we need to answer next?

NAME: ______________________

PERIOD: __________ DATE: __________

Lesson 12: What technology exists to allow us to use a form of rocket fuel in a car?

WARM UP:

1. What did we figure out in the last class? What are we wondering now?

____

____

____

____

____

REVISIT ROCKET AND INTERNAL COMBUSTION ENGINES:

2. What is similar and different about the fuels used in rocket and internal combustion engines? In what Lesson did we figure this out?

____

____

____

____

____

____

3. How is the structure and function of rock and internal combustion engines similar and different? In what Lesson did we figure this out?

____

____

____

____

RESEARCH CURRENT TECHNOLOGY:
Article 1: Where Are All the Hydrogen Cars We Were Promised?

4. What type of current technology is the engine described in the article most closely related to? ____

5. What are some advantages of this engine?

____

____

____

____

6. What are some disadvantages of this engine?

____

____

____

____

Article 2: A Hydrogen-powered Bus Goes To Washington

7. What type of engine is used to power the bus?

8. What are some advantages to this engine compared to the closely related technology?

9. What are some disadvantages to this engine?

Video: Hydrogen Cars

10. How does this technology work?

11. Thinking back to the structure and function of rocket and internal combustion engines, how is this engine's structure and function different? Similar?

NEXT STEPS:

12. What should we figure out in the next lesson?

13. What questions do you have?

Lesson 12: Engine Diagrams

Rocket Engine Diagram	Internal Combustion Engine Diagram
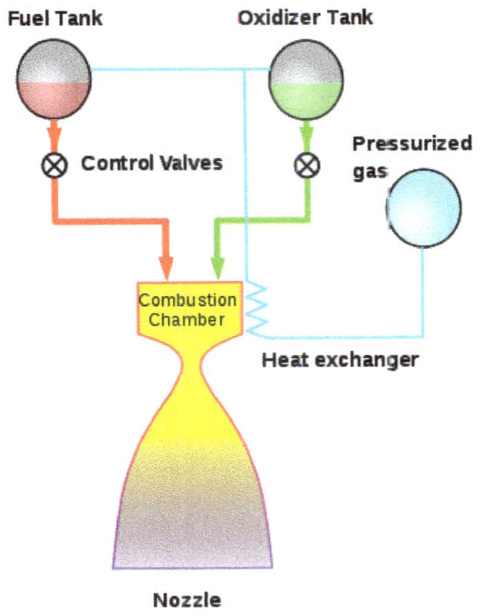	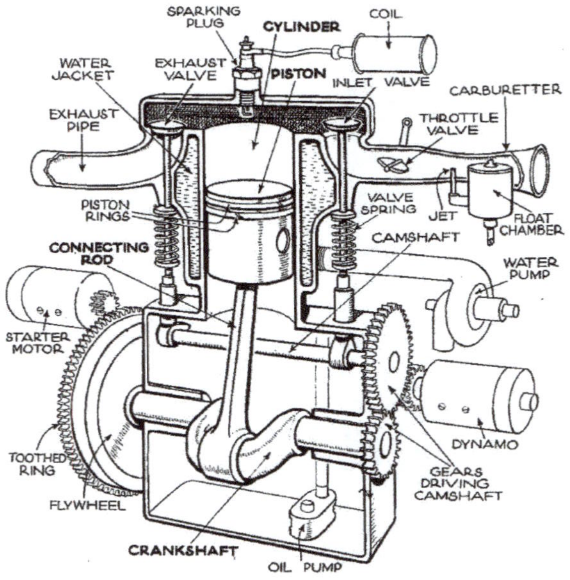
Source: Wikimedia Commons/ Duk / CC BY-SA	Source: Wkimedia Commons/ Andy Dingley (scanner) / Public Domain

NAME: ______________________________

PERIOD: ____________ DATE: ____________

Lesson 13: So...why aren't we using hydrogen fuel? Where does hydrogen come from?

LOOKING BACK

1. What did we figure out in the last class? What are we wondering now?

BRAINSTORMING

2. Use the space below to brainstorm possible reasons why we don't often use hydrogen fuel cells in automobiles.

PRIOR KNOWLEDGE

3. What do you already know about hydrogen? What did we figure out in our previous unit?

4. What might be some potential sources of hydrogen on earth?

EXAMINING DATA

5. Use the chart below to record what you notice and wonder about the data.

Notice	Wonder

PLANNING OUR INVESTIGATION

6. Use the space below to develop an investigation plan to separate the hydrogen from oxygen in water.

CARRYING OUT OUR INVESTIGATION

7. Use the space below to record data from your investigation.

RESEARCH

8. How do we get hydrogen fuel?

BRAINSTORM A LIST

9. Use the space below to brainstorm a list of criteria we need to consider when evaluating fuels.

NEXT STEPS

10. What do we need to figure out for the next class?

Lesson 13: Is Mining a Viable Option to Get Hydrogen for Fuel?

Scientists and engineers are exploring ways to make hydrogen fuel cells more available for use in personal and commercial vehicles. Since hydrogen fuel cells have zero carbon emissions, many feel this would be a way to reduce our impact on global carbon emissions. Several global manufacturers have been involved in a body called the Hydrogen Council which calculated that by 2050, hydrogen could account for 18 percent of global total final energy consumption, Figure 1.

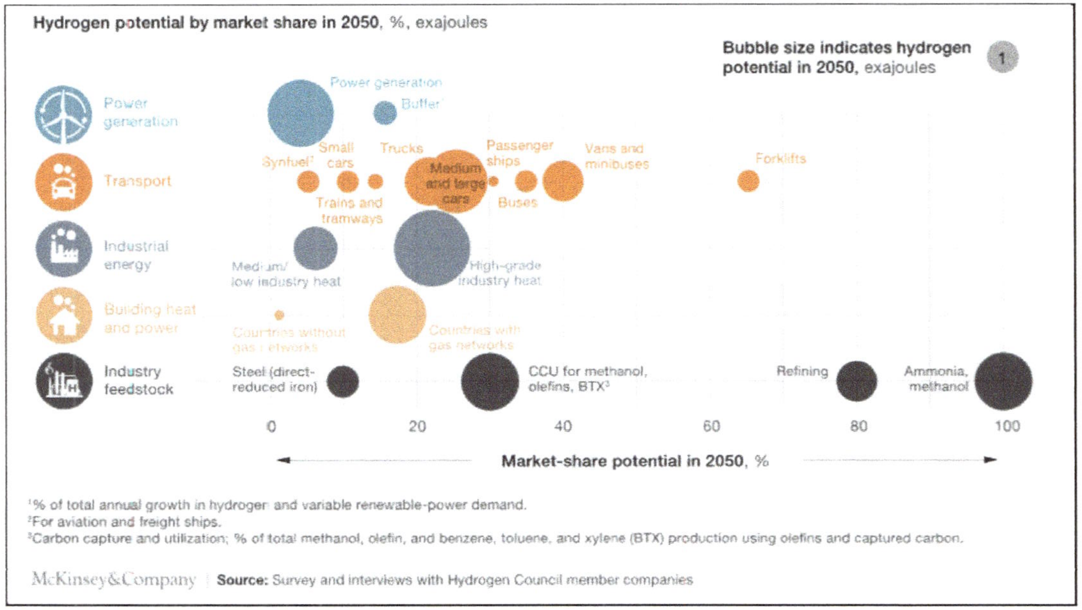

Figure 1: Percent of total annual growth in hydrogen.

Sources: https://rmi.org/a-renewable-hydrogen-way-forward-for-the-mining-industry/, https://en.wikipedia.org/wiki/Hydrogen_production, and https://www.telegraph.co.uk/china-watch/technology/hydrogen-from-water-clean-energy/

NAME: ____________________

PERIOD: __________ DATE: __________

Lesson 14: What are some possible ways to change transportation systems that would reduce or eliminate carbon emissions?

CONSENSUS DISCUSSION

1. Based on your class discussion, what claim can you make about which fuel is best? Provide evidence and reasoning from the rubric to support your claim.

RESEARCH POSSIBLE SOLUTIONS

2. Which solution did your group choose? ________________________________

Use the space below to take notes on your solution. If needed, use a separate sheet of paper for your notes.

EVALUATE A SOLUTION

3. Use the space below to take notes on your group's discussion using the weighted criteria of the solution you researched in Question 2.

4. Indicate whether your group would recommend or not recommend the solution. Provide evidence and reasoning to support your group's decision.

NEXT STEPS

5. What should we do next lesson?

____

____

____

6. What questions do you still have?

____

____

____

Lesson 14: List of Solutions for Reducing Carbon Emissions

A list of possible resources to examine different approaches to reducing carbon emissions are found below. You may choose to use an approach in this list, or research to find one of your own.

1. [Indigenous/Sovereign] https://www.indigenousclimateaction.com
2. [US national] The Green New Deal (HR 109, Vox Green New Deal Explained Article / Video)
3. [Global] The Paris Agreement
4. [Rural Community Example] Iowa Carbon Farming
5. [Hydrogen] Surprising Applications for Hydrogen Fuel Cells
6. [Large City Example] Center for Climate and Energy Solutions: City Climate Policy
7. [Activist Approaches] David Suzuki Foundation & Young Climate Activists of Color
8. [Corporate/Private Industry] Chargepoint
9. [Political Candidate Proposals] NRDC Action Fund
10. [Faith based organizations] Shiloh Temple Solar Garden

These materials were developed by the inquiryHub RPP, including educators from CO and IL, Northwestern University and the University of Colorado Boulder, with support from the Hewlett Foundation, the Spencer Foundation, the National Science Foundation, and Denver Public Schools.

NAME: ______________________________

PERIOD: ____________ DATE: ____________

Lesson 15: What should we do to decarbonize transportation and how can it be done? What part, if any, does hydrogen fuel play moving forward?

LOOKING BACK

1. What was different about each of the solutions we examined? How did they address transportation?

____

____

____

____

____

____

2. What criteria do we need to consider when evaluating different solutions? How does this impact our decisions when moving forward?

____

____

____

____

____

____

INITIAL IDEAS

3. Why do you think hydrogen fuel hasn't been used on a larger scale to reduce carbon emissions by cars?

____

____

____

____

____

RESEARCH WAYS TO COMMUNICATE

4. Use the space below to take notes over your section of the reading.

Section Title: ______________________________________

5. What are the key ideas of your section?

The Psychology of Climate Change Communication

A Guide for Scientists, Journalists, Educators, Political Aides, and the Interested Public

Center for Research on Environmental Decisions

The Psychology of Climate Change Communication

A Guide for Scientists, Journalists, Educators, Political Aides, and the Interested Public

CRED.COLUMBIA.EDU/GUIDE

CRED Center for Research on Environmental Decisions

ACKNOWLEDGEMENTS

Writers
Debika Shome and Sabine Marx

With contributions from the following members of CRED
Kirstin Appelt, Poonam Arora, Roberta Balstad, Kenny Broad, Andrew Freedman, Michel Handgraaf, David Hardisty, David Krantz, Anthony Leiserowitz, Massimo LoBuglio, Jenn Logg, Anna Mazhirov, Kerry Milch, Nancy Nawi, Nicole Peterson, Adrian Soghoian, Elke Weber

Reviewers
Mary-Elena Carr, Kelly Hayes-Raitt, Betsy Ness-Edelstein, Clare Oh, Gavin Schmidt, Renzo Taddei, Bud Ward

Editor
Andria Cimino, Leapfrog Communications

Illustrator
Ian Webster, Hazard County Illustration

Graphic Designer
Erich Nagler, Design Means

Special Thanks
The Earth Institute, Columbia Climate Center, The Harmony Institute, the Leonard and Jayne Abess Center for Ecosystem Science and Policy, Rosenstiel School of Marine and Atmospheric Science, University of Miami, and National Geographic Society.

This guide was made possible by the generous support of:
The Charles Evans Hughes Memorial Foundation
National Science Foundation SES-0345840

CGP-EGC/PR-1001

This book was printed with a Certified Green Partner, ensuring that the paper contains fibers from sustainable and well-managed forests, and the use of vegetable-based inks.

Citation: Center for Research on Environmental Decisions. (2009). *The Psychology of Climate Change Communication: A Guide for Scientists, Journalists, Educators, Political Aides, and the Interested Public*. New York.

For an online version of this guide, visit cred.columbia.edu/guide.

About the Center for Research on Environmental Decisions
CRED is an interdisciplinary center that studies individual and group decision making under climate uncertainty and decision making in the face of environmental risk. CRED's objectives address the human responses to climate change and climate variability as well as improved communication and increased use of scientific information on climate variability and change. Located at Columbia University, CRED is affiliated with The Earth Institute and the Institute for Social and Economic Research and Policy (ISERP). For more information, visit cred.columbia.edu.

We welcome feedback on this guide. Please send emails to cred@columbia.edu.

CONTENTS

Introduction

"The ultimate solutions to climate change are workable, cost-effective technologies which permit society to improve living standards while limiting and adapting to changes in the climate. Yet scientific, engineering, and organizational solutions are not enough. Societies must be motivated and empowered to adopt the needed changes.

For that, the public must be able to interpret and respond to often bewildering scientific, technological, and economic information. Social psychologists are aware, through their painstaking scientific research, of the difficulties that individuals and groups have in processing and responding effectively to the information surrounding long-term and complex societal challenges.

This guide powerfully details many of the biases and barriers to scientific communication and information processing. It offers a tool—in combination with rigorous science, innovative engineering, and effective policy design—to help our societies take the pivotal actions needed to respond with urgency and accuracy to one of the greatest challenges ever faced by humanity: global-scale, human-induced environmental threats, of which the most complex and far reaching is climate change."

—Jeffrey Sachs, Director, The Earth Institute, Columbia University

WHY AREN'T PEOPLE MORE CONCERNED ABOUT CLIMATE CHANGE?

Research shows that most Americans do not feel a personal connection to climate change.[1] They are aware of it, they may even rank it as a concern, but according to a 2008 Pew Research Center for People and the Press, they do not perceive it as a near-term priority on par with, say, the economic downturn or the need to reform health care. In fact, despite scientists' calls for urgent action, climate change has slipped to the bottom of the list of American priorities.[2]

Many people can recite at least a few things they could do to help mitigate global climate change, but are not. Why not? Somehow, and despite a lot of media attention following the release of *An Inconvenient Truth*, messages about climate change and what people need to do to help prevent it seem to have fallen on deaf ears.

There are many theories about why awareness of climate change does not inspire the kind of behavior changes it should. Addressing all of them goes beyond the scope of this guide. What this guide does provide are principles derived from the social sciences concerning how to communicate effectively about a topic that is complex, confusing, uncertain, sometimes overwhelming, and often emotionally and politically loaded.

CRED research shows that, in order for climate science information to be fully absorbed by audiences, it must be actively communicated with appropriate language, metaphor, and analogy; combined with narrative storytelling; made vivid through visual imagery and experiential scenarios; balanced with scientific information; and delivered by trusted messengers in group settings. This guide combines laboratory and field research with real-world examples. It blends information from the broad spectrum of disciplines that CRED encompasses: psychology, anthropology, economics, history, environmental science and policy, and climate science.

Intended for anyone who communicates about climate change, from scientists, journalists, educators, clerics, and political aides to concerned citizens, the guide's purpose is to assist communicators in reaching two key audiences—the general public and decision makers from government and business—more effectively. The principles found in this guide should help make climate change presentations and discussions more effective.

TERMINOLOGY NOTE

Climate Change vs. Global Warming

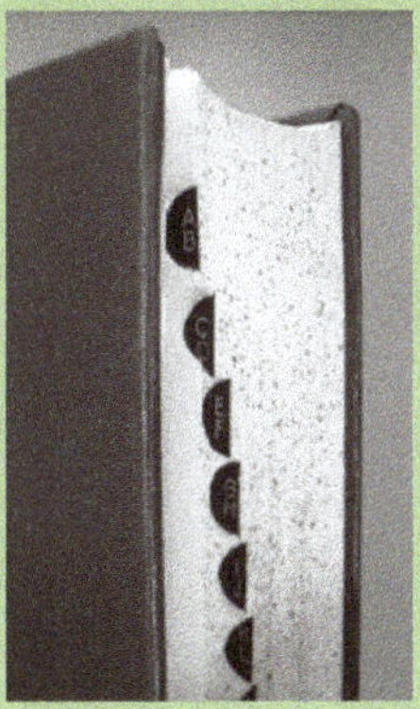

Erich Nagler

This guide uses the term *climate change* to refer to the changes that are occurring in the earth's climate system and the impacts such changes are having on ecosystems and society. *Climate change* is a better choice than the term *global warming* because it avoids the misleading implications that every region of the world is warming uniformly and that the only dangerous outcome of growing greenhouse gas emissions is higher temperatures, when that, in fact, is just the tipping point for a cascade of changes in the earth's ecosystems.

In addition, *climate change* better conveys the coexistence of human-made effects with natural climate variability, a more accurate, "state-of-the-science" portrayal of the causes for the phenomenon. Since the climate is warming in fits and starts rather than on a constant basis, each year might not be warmer than the previous one. As this guide will show, using more precise terminology (and defining easily misunderstood terms) is a vital ingredient to clear, memorable, and impactful communication.

1 Know Your Audience

WHAT IS A MENTAL MODEL?

A **mental model** represents a person's thought process for how something works (i.e., a person's understanding of the surrounding world). Mental models, which are based on often-incomplete facts, past experiences, and even intuitive perceptions, help shape actions and behavior, influence what people pay attention to in complicated situations, and define how people approach and solve problems.[3] Perhaps most important to climate change communicators, mental models serve as the framework into which people fit new information.[4]

People usually have some relevant knowledge and beliefs that help them interpret new information in order to reach conclusions. When hearing about risk, people often refer to known related phenomena and associations from their past to decide if they find the risk threatening or manageable. But sometimes a mental model serves as a filter, resulting in selective knowledge "uptake," i.e., people seek out or absorb only the information that matches their mental model, confirming what they already believe about an issue. This poses a potential stumbling block for climate change communicators.

MENTAL MODELS AND THE CONFIRMATION BIAS

A **confirmation bias** makes people look for information that is consistent with what they already think, want, or feel, leading them to avoid, dismiss, or forget information that will require them to change their minds and, quite possibly, their behavior. People often exhibit a strong preference for their existing mental models about climate change, making them susceptible to confirmation biases that lead them to misinterpret scientific data, as shown by the example below.

HOW TO IDENTIFY AND UPDATE MENTAL MODELS ABOUT CLIMATE CHANGE

The good news is that mental models are not static—people will update them by correcting misinformation, inserting new building blocks, and/or making new connections with existing knowledge. But for a presentation of new climate change information to succeed, communicators should first do their best to discover what climate change misconceptions the audience may have in its mental models. Communicators can then disconnect the erroneous climate change information from other parts of the model and replace it with new facts. The example on page 5 explores a common misconception that climate change communicators run into and how to counter it.

EXAMPLE

The Confirmation Bias and Climate Change

Both believers and skeptics find it tempting to over-interpret short-term hot or cold swings in temperature as evidence for or against climate change. Such confirmation bias in action can lead people who believe that climate change equals warmer temperatures to pay greater attention to supportive data, interpreting a heat wave in the Great Lakes region, for example, as evidence that their mental model is correct. Skeptics of climate change might pay more attention to news that announces close-to-normal levels of polar sea ice, a momentary finding but one that fits their mental model and enables them to disregard the more scientifically relevant trend of dramatic loss of sea ice in the Arctic and "debunk" climate change.[8] Both sides will either ignore facts that contradict their mental model of climate change or interpret them as exceptions to the rule.

Erich Nagler

Dr. Gavin Schmidt of NASA's Goddard Institute for Space Studies discussed the phenomenon of trying to fit new information into people's existing beliefs about climate change, providing communicators with great advice. As he explained to the *New York Times,* "there is this desire to explain everything that we see in terms of something you think you understand, whether that's the next ice age coming or climate change…When I get called by CNN to comment on a big summer storm or a drought or something, I give the same answer I give a guy who asks about a blizzard. 'It's all in the long-term trends. Weather isn't going to go away because of climate change.'"[6]

EXAMPLE

A Common Mental Model about the Ozone Layer and Greenhouse Gases

CRED researcher and director of the Yale Project on Climate Change Anthony Leiserowitz finds that people often confuse the hole in the ozone layer with climate change.[7] This is happening, in part and ironically, due to a science communication victory. Scientists and the media effectively and extensively covered the threat posed by the growing ozone hole, eventually resulting in international political action to phase out the main contributor, chlorofluorocarbons or CFCs.

But now many people conflate their mental model of the ozone layer with how the atmosphere works, in particular with how greenhouse gases accumulate in the atmosphere. Dr. Leiserowitz has found it leads to some interesting misconceptions that require updating, such as: If there is a "hole" in the ozone layer, and there is a global "greenhouse" effect, then there must be a "hole" in the "greenhouse." Some Americans thus reason that this "hole" either allows more solar radiation into the biosphere—warming the planet—or, alternatively, allows heat to escape—cooling the planet.

Although logical, such reasoning has unfortunately led to construction of an inaccurate mental model about the causes of climate change that, in turn, causes many Americans to support inappropriate solutions, such as believing that the best way to solve global warming is to ban aerosol spray cans.[8] Climate change communicators should try to identify this commonly mistaken mental model and replace it with correct information.

2 Get Your Audience's Attention

WHAT IS FRAMING?

Framing is the setting of an issue within an appropriate context to achieve a desired interpretation or perspective. The intention is not to deceive or manipulate people, but to make credible climate science more accessible to the public. Indeed, since it is impossible **not** to frame an issue, climate change communicators need to ensure they consciously select a frame that will resonate with their audience.

WHY FRAME?

Below are just a few of the benefits derived from framing:[9]

- Frames organize central ideas on an issue. They endow certain dimensions of a complex topic with greater apparent relevance, more so than the same dimensions might appear to have under an alternative frame.
- Frames can help communicate why an issue might be a problem, who or what might be responsible, and, in some cases, what should be done.
- Frames can help condense a message into useful communication "short cuts" and symbols: catch-phrases, slogans, historical references, cartoons, and images.

THE SUBTLETIES OF FRAMING

Communicators should keep in mind both form and content when framing a climate change message. The first step, as explained in Section 1, is determining as much as possible about an audience's mental models of climate change.

The next step is to consider the audience's membership in specific subcultures (groups of people with distinct sets of beliefs, or based on race, ethnicity, class, age, gender, religion, occupation). Is there a majority represented in the audience? For instance, will college students concerned about the creation of green jobs comprise the audience? Or city officials interested in increasing energy efficiency standards in building codes? Consider the local perspective—an audience in Colorado may identify more strongly with the links between climate change and threats to the ski industry, whereas a group from Florida may care more about the links between climate change and sea level rise.

Communicators may find it useful to prepare numerous frames ahead of time, including climate change as a religious, youth, or economic issue. They should be

EXAMPLE

Framing and the Politics of Carbon

Although many top economists and climate scientists agree that a carbon tax would be the most streamlined step the US could take to reduce its contribution to climate change, support for a carbon tax among major politicians is extremely limited.[11] Yet, at the same time, many businesses and individuals voluntarily purchase "carbon offsets" (or "carbon credits"), which promise to balance out the greenhouse gases produced by a particular activity they are engaging in. How much of this support is a reflection of the framing power of the words *tax* and *offset*?

Martin Jacobsen, commons.wikimedia.org

CRED researchers polled a large national sample about a program that would raise the cost of certain products believed to contribute significantly to climate change (such as air travel and electricity) and use the money to fund alternative energy and carbon capture projects.[12] The identical program was described as a "carbon tax" to half the respondents, and as a "carbon offset" to the other half. This simple change in frame had a large impact on people's preferences.

When considering a pair of products, 52% of respondents said they would choose a more expensive product when the cost increase was labeled a "carbon offset," but only 39% when it was labeled a "tax." Support for regulation to make the cost increase mandatory was greater when it was labeled an "offset" than when it was labeled a "tax."

Strikingly, the framing effect interacted with respondents' political affiliations. More liberal individuals did not discriminate between the two frames (meaning, they were equally likely to support the program regardless of the label used), but more conservative individuals strongly preferred the carbon offset to the carbon tax. A follow-up study revealed that the tax label triggered many negative thoughts and associations among more conservative individuals, which in turn led them to reject the carbon tax.

careful when framing climate change, however, not to focus so intently on one particular aspect that the audience loses sight of the bigger picture.

When dealing with scientific communication and framing, research shows that an audience may be more receptive when they perceive the information being communicated as having salience, relevance, authority and legitimacy.[10] As the example on page 7 illustrates, framing can be a subtle art—even the choice of a single word can make the difference between winning and alienating an audience.

PROMOTION VS. PREVENTION FRAME

A powerful way of framing a message considers people's goals. Do they view their goal as making something good happen, or preventing something bad from happening?

People approach goals differently. People with a **promotion focus** see a goal as an ideal and are concerned with advancement. They prefer to act eagerly to maximize or increase gains. People with a **prevention focus**, however, see a goal as something they ought to do and are concerned with maintaining the status quo. They prefer to act vigilantly to minimize or decrease losses.[13]

Research shows that tailoring messages to people's natural promotion and prevention orientations increases the level of response for both groups, regardless of whether their response was positive or negative. These findings support the idea of framing messages from multiple perspectives to accomplish environmental goals. For example, if a local city wants people to increase their recycling, city officials should explain options in different ways, some with a promotion focus and some with a prevention focus. A promotion message would emphasize "going the extra mile" (e.g., going out of one's way to recycle, how recycling benefits the community). A prevention message would encourage "dotting the 'i' and crossing the 't'" (e.g., being careful to recycle, how not recycling hurts the community).

When communicators craft their climate change messages, they should remember that framing requires the careful selection of words that will resonate with the audience's orientations. The table to the left lists words that will help communicators frame messages that appeal to those who are promotion focused and/or prevention focused. In order to increase the chances of reaching a greater number of people in the audience, communicators should include both promotion- and

TABLE 1

Words That Appeal To Those with Either a Promotion or Prevention Focus[14]

PROMOTION	PREVENTION
ideal	ought
attain(ment)	maintenance
maximize gains	minimize losses
hope	responsibility
wish	necessity
advance(ment)	protect(ion)
eager(ness)	vigilant/vigilance
avoid missed opportunities	avoiding mistakes
promote	prevent
aspire/aspiration	duty
support	obligation
nurture	defend
add	safety
open	security
	must
	should
	cautious
	careful
	stop

prevention-oriented wording in their messages. People feel better and more positive about achieving their goals and are more likely to sustain their behavior when their goals are framed in a manner that feels naturally comfortable to them.

BRING THE MESSAGE CLOSE TO HOME: A LOCAL FRAME

Although the majority of US residents consider climate change a serious problem, they generally think of it in geographically and temporally distant terms. Most US residents do not personally experience effects that are drastic enough on a regular basis to alarm them about climate change.

In a July 2007 national survey, respondents believed that climate change was a "very serious threat" for "plants and animals" (52%), "people in other countries" (40%) and "people elsewhere in the United States" (30%). However, far fewer saw it as a "very serious threat" to "you and your family" (19%) and "your community" (18%).[15] In other words, people perceive climate change impacts as a threat to plants and animals and people in other parts of the world, but do not see it as a local issue affecting themselves, their family, and their community.

To counteract this problem, an effective communicator should highlight the current impacts of climate change on regions within the US. Research suggests that it may be more effective to frame climate change with local examples in addition to national examples. For example, references to droughts in the Southwest may resonate more with US audiences than talking about droughts in Africa. Similarly, climate change becomes a more personal threat to a New Yorker when hearing how New York City's subway system will suffer as the result of a rise in sea level compared to hearing about the effect of a sea level rise in Bangladesh.

Scientists have found trends in extreme weather events, such as heat waves and flooding, that are consistent with climate change, such as increasing precipitation extremes in some parts of the United States.[16] They project that climate change will increase the frequency and/or severity of extreme events, which may in turn cause significant damage to human health and infrastructure.

Because such extreme weather events are vivid, dramatic, and easily understood, especially to the locals who suffer through them, they provide effective frames for the potential impacts of future climate change. The numerous examples of extreme events that may occur in a given year provide recurring "teachable moments" communicators can use to relate climate change to the experience of a local audience. For example, the long-lasting and severe drought, known as the "Big Dry," that has gripped some parts of Australia since 1995 has increased that public's awareness of climate change, resulting in greater support for measures to combat climate change.[17]

When discussing extreme events, however, communicators should bear in mind that while it is correct to say that climate change is increasing the odds of an extreme weather event, climate scientists cannot yet make the claim that climate change is **causing** such events. This important distinction often gets lost or is misunderstood, causing confusion and undue skepticism (see Section 5 for more about how to address scientific uncertainty).

For example, although scientists can link the severity of Hurricane Katrina to increased warming in the ocean, it is difficult to distinguish how much of that warming is due to human activity and how much is the result of a natural cycle.[18] Therefore, it is incorrect to say that climate change caused Katrina. Communicators should also be careful not to conflate cause and effect; a variety of factors conspired to make the aftermath of Hurricane Katrina so damaging and deadly. Katrina did, however, provide a powerful example of how costly extreme weather events can be, even if they aren't the direct result of climate change.

Communicators can also use local frames to effectively illustrate climate change solutions. In Montana's Glacier National Park, where climate change has led to the loss of numerous glaciers, the state government has begun investing in new "green" technologies like carbon sequestration and windmills.[19]

By framing climate change as a local issue, communicators not only increase their audience's sense of connection to and understanding of climate change, but also promote the development of local and regional solutions that could transfer well to the national and global arenas and, further, inspire future action everywhere.

MAKE THE MESSAGE MATTER NOW: THE NOW VS. FUTURE FRAME

People typically perceive immediate threats as more relevant and of greater urgency than future problems.[20] Yet communicators often portray the threat from climate change as a future rather than present risk.

The problem with this approach is that people tend to discount the importance of future events. Indeed, many social scientists believe that this is one of the top reasons that it is hard to motivate people to take action to prevent climate change. CRED research documents that many people count environmental and financial consequences as less important with every year they are delayed.[21] For example, the average person finds little difference between getting $250 now or $366 in one year (implying an interest rate of roughly 46%). The rates of discounting are similar for environmental consequences; the average person finds little difference in 21 days of clean air now over 35 days of clean air next year. Fortunately, communicators can make this predisposition (to heavily discount future larger losses) work to help people overcome a reluctance to take on immediate losses.

For example, if a communicator wants audience members to sign up for weatherizing their homes (which increases a home's energy efficiency), he or she may have more success by having them commit to an evaluation of their home's efficiency three or six months into the future rather than immediately. Because future consequences are discounted, people often think, "I'm busy now, but in the future I'll have more time and it won't be such a big deal." Of course, the key is to ensure audience members make a binding agreement so that the evaluation and subsequent weatherization really happen in the near future. In this case, a communicator might want to have a sign-up sheet that will enable audience members to sign up for a specific appointment with a local company that offers this service or for a time the local power company might contact them to schedule an appointment.

COMBINING THE NOW VS. FUTURE FRAME WITH A GAIN VS. LOSS FRAME

The negative feelings associated with losing $100 outweigh the positive feelings associated with gaining $100. Thus people have a natural tendency to avoid losses rather than to seek gains.[22] When a gain vs. loss frame is combined with a now vs. future frame, people discount future gains more than future losses.[23] For example, people may be more likely to adopt environmentally responsible behavior and support costly emissions reduction efforts related to climate change if they believe their way of life is threatened and that inaction will result in even greater loss. They are less likely to adopt these measures if they focus on the current situation which they see as acceptable and discount future improvement of it.

It may be possible to motivate environmentally responsible behavior by tapping into people's desire to avoid future losses rather than realize future gains. For instance, when communicators talk to homeowners, they could frame energy efficiency appliances as helping the homeowners to avoid losing money on higher energy bills in the future, instead of helping them save money in the future. Campaigns to encourage people to buy fuel-efficient vehicles could focus on how their use will avoid continuing and even increasing future losses in money to pay for gasoline instead of how such cars will save the consumer money.

To hold an audience's attention and encourage behavior change, communicators may want to present information in a way that makes the audience aware of potential current and future losses related to inaction on climate change instead of focusing on current and future gains. Audiences may be more likely to make changes to their behavior if climate change information is framed as 'losing less now instead of losing more in the future'. For example, during hot summer months, a smaller reduction in daily energy use can prevent having to deal with larger and prolonged energy blackouts.

BROADEN THE MESSAGE: THE INTERCONNECTED FRAME

"Our nation has both an obligation and self-interest in facing head-on the serious environmental, economic, and national security threats posed by climate change."
—Senator John McCain (R-Arizona)

Communicators sometimes frame climate change as if it is only an environmental problem, which enables some people to shrug it off as something only environmentalists need to worry about. But climate change is not a disconnected phenomenon that will only affect the weather and ecosystems, but a change that will have an impact on nearly every human system, including health, the economy, and national security. Vice President Al Gore hinted at the interconnected nature of climate change when he described it in his 2007 Nobel Peace Prize acceptance speech as "real, rising, imminent, and universal."[24]

Below and on page 13 are two examples of how communicators can broaden their messages to include such frames as national security and human health.

EXAMPLE

The National Security Frame

"We will pay for this [climate change] one way or another. We will pay to reduce greenhouse gas emissions today and we'll have to take an economic hit of some kind. Or we will pay the price later in military terms. And that will involve human lives."
—Gen. Anthony C. Zinni, retired Marine and former head of the Central Command

"The traditional triggers of conflict which exist out there are likely to be exacerbated by the effect of climate change."
—Emyr Jones Parry, Great Britain's ambassador to the United Nations

When it comes to national security threats, policymakers usually place terrorism as their main concern, with global climate change ranking far below, if at all. Recently, however, climate change has secured its place as a national security issue on both a national and global scale. In 2007, for example, the United Nations Security Council put climate change on its agenda for the first time, warning that it could serve as a catalyst for new conflicts around the world.[25] National security concerns deriving from climate change include the reduction of global food supplies, leading to large migrations of populations; increased risks for infectious disease, including pandemics that could destabilize economies and governments; and increased fighting over already limited resources like water and land.[26]

When talking about climate change, communicators should frame their messages to match what they think the audience may already relate to and worry about in terms of national security. For instance, when speaking to people in the military, communicators could highlight the connections between climate change and potential conflicts over natural resources, especially by so-called "failed states," a term often used to describe a state perceived as having failed at some of the basic conditions and responsibilities of a sovereign government. When speaking with a group of parents, communicators might want to describe what the world could be like when their children are adults—when issues like water wars, food shortages, and sheltering environmental refugees may become realities for people in the US.

EXAMPLE

The Human Health Frame

US residents may not perceive climate change as a threat to human health, either now or in the future. Some of the health implications related to climate change are relatively well understood (e.g., an increased likelihood of heatstroke), while others are less obvious (e.g., the rapidly rising rates of asthma and respiratory conditions). Drawing awareness to the connections between climate change and human health may be an effective method for elevating public concern about climate change in the US. By articulating its serious individual health consequences, communicators can help frame climate change as a concrete, personal concern for everyone.

Framing solutions, such as developing alternative energy, in terms of health and climate change also works well. The burning of fossil fuels like coal and oil creates "dirty energy" and emits large amounts of gases, such as carbon dioxide and nitrogen oxide. In addition to advancing climate change, these emissions directly pollute air and water by forming particulates like soot. It may be beneficial to highlight the multiple positive effects of reducing the burning of fossil fuels. Since direct or indirect exposure to these pollutants can cause cancer, breathing problems, birth defects, and mental impairment, reducing climate change-related pollution would also help reduce negative health outcomes.[27]

3 Translate Scientific Data into Concrete Experience

The famous "Keeling curve" graph, below, which shows the increasing amounts of carbon dioxide in the earth's atmosphere from 1958 to 2006, set off alarms in the scientific community that continue to ring loudly today. Yet somehow, this same graph does not communicate the immediacy of the climate change problem to lay audiences. Instead, it may actually convey the message that the buildup of carbon dioxide in the earth's atmosphere has been taking place over a long period, thereby erroneously implying that climate change is not an urgent issue.

Similarly, many people have difficulty grasping the importance of projections of higher carbon dioxide concentrations and surface temperatures several decades from now. Part of the problem may be the tendency to discount future events, as described in Section 2. But another part of the problem may be that a global average surface temperature increase of a few degrees does not seem like much to the general public, given the variability in temperature

THE "KEELING CURVE"

Mauna Loa Observatory, Hawaii
Monthly Average Carbon Dioxide Concentration

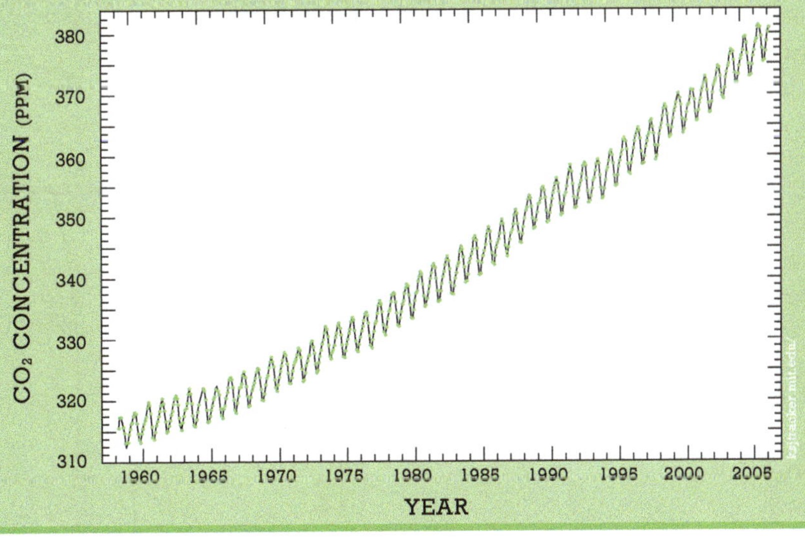

that most people experience on a regular basis.

But a few degrees do matter. As the 2007 United Nations Intergovernmental Panel on Climate Change (IPCC) report found, numerous effects of climate change are already observable throughout the earth system, and these impacts are likely to grow in coming years.[28] Yet polls taken during the past several years continue to show Americans ranking climate change near the bottom of their list of concerns or policy priorities.[29] Clearly attempts to convey the immediacy of the climate challenge have fallen short of translating climate change into a near-term (as well as a long-term) danger on par with other imminent societal and personal threats.

WHY THE "KEELING CURVE" ALONE DOESN'T MOTIVATE BEHAVIOR CHANGE

Many of the highly publicized graphs and charts showing global climate change data pose a problem for communicators because they fail to inspire a sense of urgency in many audiences. They do not help convey the deep concern scientists have that efforts to abate and adapt to climate change are a near-term necessity if humanity is to avert the worst effects. Despite making this point with increasing frequency and stronger data, the general public shows little concern.

Even when people understand the Keeling Curve, it does not always motivate them to take action. The reason for this disconnect may lie in how the brain works, which climate change communicators need to understand to create truly powerful messages that will inspire action.

HOW THE BRAIN PROCESSES INFORMATION

The human mind is not designed to immediately react to threats that seem to manifest themselves in the distant future, such as climate change. Distant risks do not set off the same alarms that immediate risks do. Human brains struggle to balance long-range worries with the demands of more immediate concerns.[30]

More specifically, the human brain has two different processing systems: the experiential processing system, which controls survival behavior and is the source of emotions and instincts (e.g., feeding, fighting, fleeing); and the analytical processing system, which controls analysis of scientific information. Table 2 on page 16 highlights the key differences between these two systems.

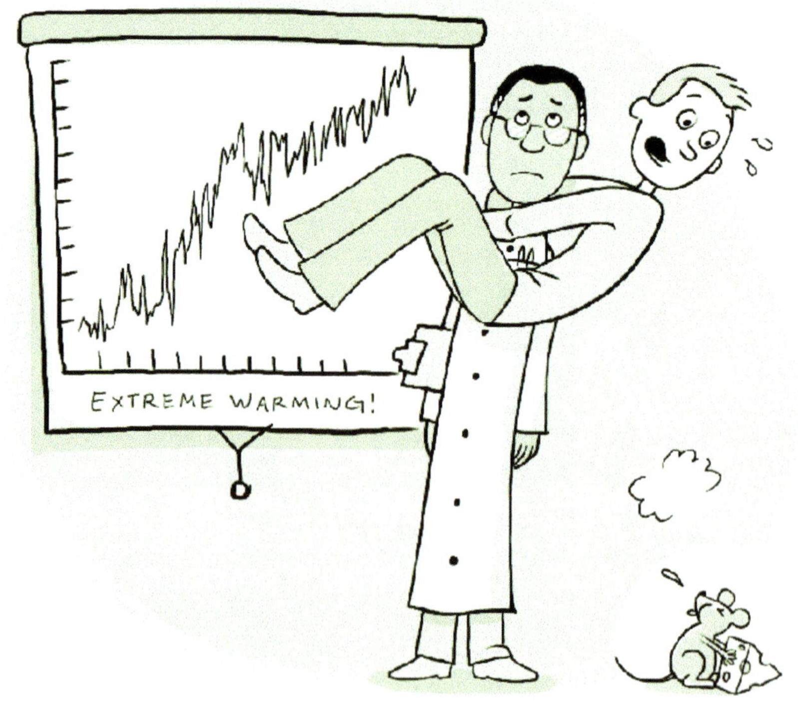

TABLE 2

Two Information Processing Systems of the Brain[31]

ANALYTIC PROCESSING SYSTEM	EXPERIENTIAL PROCESSING SYSTEM
Logical	Holistic
Deliberative	Intuitive
Analytic	Emotion-driven (fear, dread, anxiety)
Perceives reality in abstract symbols, words, numbers	Perceives reality in concrete images and narratives, linked in associations
Rules and algorithms need to be learned; system needs to be prompted; does not operate automatically	Operates automatically and without any training
Examples • numerical statistics in tables, figures, graphs, charts	**Examples** • images or stories • the experience of outcomes of repeated decisions over time, as in a simulation exercise • emotionally charged and vivid

SPEAK TO THE TWO PARTS OF THE BRAIN: HOW TO MAKE ANALYTIC DATA MEMORABLE AND IMPACTFUL

Traditional statistical presentations of climate change data rarely instill the sense that it is an immediate challenge as well as a future one; that there is a narrow window of opportunity within which effective action can avert potentially devastating future consequences. Many audiences leave such analytically focused presentations with a higher awareness that climate change is happening, but without the matching higher motivation to do anything about it.

Despite evidence from the social sciences that the experiential processing system is the stronger motivator for action, most climate change communication remains geared toward the analytical processing system. Personal or anecdotal accounts of negative climate change experiences, which could easily outweigh statistical evidence, are rarely put into play, despite evidence that even a stranger's past experiences can evoke strong feelings in people, making such communications memorable and therefore dominant in processing.[32]

Yet not all communication about climate change should be emotional, as there are downsides to bypassing analytical reasoning to make an appeal only to the experiential system (Section 4 will address these climate change communication pitfalls).

3

Translate Scientific Data into Concrete Experience

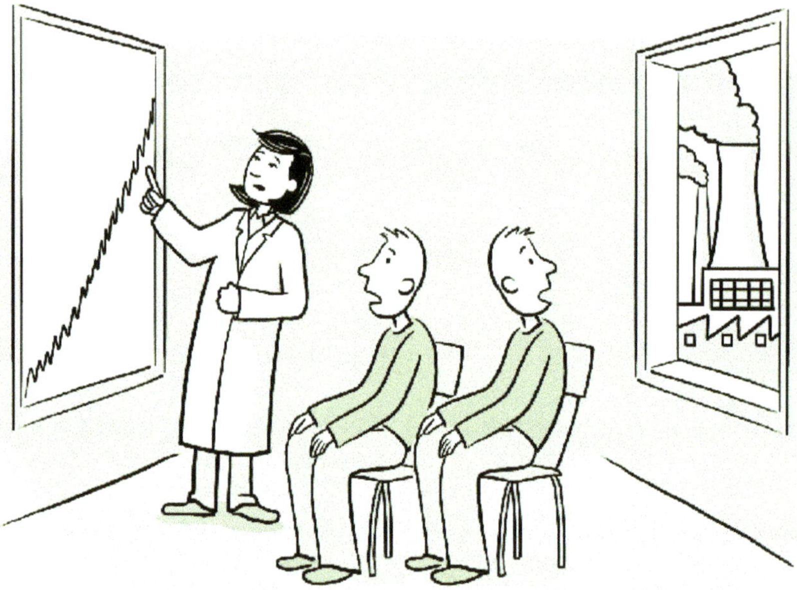

EXAMPLE

Shrinking Glaciers and the Retention of Facts

In 2007, CRED researchers developed an interactive computer presentation to show viewers the effect of climate change on the world's glaciers. One module presented information that would appeal to the analytical processing system, such as scientific analysis, statistics, and graphs, to describe the relationship between climate change and shrinking glaciers. Another module targeted the experiential processing system of the brain, using vivid imagery (photographs, videos showing reduced glacier size over time, local news footage) and personal accounts to convey the message. After randomly viewing either the analytic or experiential module on shrinking glaciers, students took a survey that measured their environmental attitudes, perceptions, and behaviors. The purpose was to test the module's effect on memory and the students' decision-making processes. The learning modules examined the extent to which experience-based vs. analytically framed information influenced feelings of worry, risk perception, and the willingness to take action about climate change.

L. Chang, commons.wikimedia.org

The results showed that people retained more factual information about the presentation after viewing the experiential module as compared to the analytic format. CRED also found that when students viewed the experiential module, they reported both increased levels of worry and willingness to take action.[33]

Unfortunately, the resulting willingness to take action after an appeal to the experiential processing system alone can be short-lived. Section 4 will explore why emotional appeals about climate change can backfire and how to avoid this phenomenon.

EXAMPLE

The Effect of Vivid Imagery on Recycling in New York City

© Katvan Studios, courtesy NYC & Company/OROE

In 2008, the City of New York and partners launched an advertising campaign to promote recycling awareness in the city. Recognizing that the average American may have difficulty processing information if it is presented in a strictly technical format, the campaign used metaphor, analogy, vivid imagery, and comparison to communicate facts such as: "New York City residents discard 800 million pounds of recyclable paper annually." The ads powerfully illuminated how the huge amount of recyclable paper thrown away in New York City annually is enough to fill the entire Empire State Building by creating a picture of the iconic skyscraper composed entirely of discarded magazines and catalogs.

Unlike the more typical city-issued recycling advisory, this ad helped grab the viewer's attention and personalized the message in order to encourage people to change their behaviors.

The most effective communication targets both processing systems of the human brain. Communicators should make use of the following experiential tools in addition to the more common analytical ones when creating presentations on climate change:

- Vivid imagery, in the form of film footage, metaphors, personal accounts, real-world analogies, and concrete comparisons;
- Messages designed to create, recall, and highlight relevant personal experience and to elicit an emotional response.

Analytic products (such as trend analyses, forecast probabilities, and ranges of uncertainty) help people absorb facts and can be valuable tools when people need to make big decisions, but they alone will not compel people to take effective steps to address the climate change challenge, as the example on page 17 illustrates.

The example above shows how information balanced with both analytic and experiential materials may be more likely to have an effect on attitudes and behavior, creating a desire in people to act on their new knowledge.

USE UNDERSTANDABLE LANGUAGE

Another possible reason for the public's lack of responsiveness to climate change messages may be caused by low comprehension of or interest in communications laden with scientific language. When talking to the general public, research shows that communicators should, whenever possible, avoid using jargon, complicated scientific terms, and acronyms. Instead, use words that will make sense to the audience.

Table 3 below contains words or phrases that are commonly used when discussing climate change and alternative words that get the same idea across more simply.

Sometimes only a scientific term is sufficient for getting a point across. In that case, it is important to thoroughly define the term for the audience. Communicators should remember, however, that stringing together too many scientific terms and acronyms may cause the audience to spend their time and mental energy deciphering vocabulary instead of absorbing the overall point.

TABLE 3

Examples of Simplified Scientific Terms

OBSCURE WORD	BETTER UNDERSTOOD WORD
Anthropogenic	Human induced, man-made
Mid-Pleistocene	1 million to 600,000 years ago
CH4	Methane
IPCC	The group of scientists who issue comprehensive assessments on climate science, and were awarded the 2007 Nobel Peace Prize for their work on climate change.
Forcing	Incoming and outgoing (radiation) energy
385 ppm	2008 level of carbon dioxide in the atmosphere
Bifurcation	To divide into two parts
Perturbation	Disturbance
Aerosol	Small atmospheric particle

4 Beware the Overuse of Emotional Appeals

It may be tempting to conclude that an effective way to communicate climate change information is to place a greater emphasis on its possible consequences. Some go even further, accentuating the risks by declining to mention the uncertainties involved. Such an approach evokes strong reactions in audiences, including fear of worst-case climate change scenarios and even heightened interest in what can be done to avoid them. But while an emotional appeal may make people more interested in a presentation on climate change in the short run, it may backfire down the road, causing negative consequences that often prove quite difficult to reverse.

WHAT IS THE FINITE POOL OF WORRY?

Researchers at CRED and elsewhere have discovered that people, even those who might be described as "worrywarts," have a limited capacity for worrying about issues. Scholars refer to this limited capacity as a **finite pool of worry**,[34] and it has three main components that apply to the issue of climate change:

1. Because people have a limited capacity for how many issues they can worry about at once, as worry increases about one type of risk, concern about other risks may lessen. In other words, people tend to pay more attention to near-term threats, which loom larger than long-term ones.[35] For example, as anxiety mounted in 2008 and 2009 over the faltering economy, polls showed that many people realigned their list of concerns. The economy vaulted to the top of the list, while environmental issues and climate change fell to the bottom. A recent poll showed that climate change ranked last among the public's list of top policy priorities.[36]

   In another example, farmers in Argentina were asked to rate how much they worried about political risks, weather and climatic risk, and economic risks. Then farmers were shown a climate forecast for the following spring, predicting less rain than normal. As expected, farmers perceived climate as a greater risk after they had been shown the forecast. Yet, as the concern about climatic risk increased, concern about political uncertainty diminished, even though the political risk had not changed.[37]

2. Studies show that appeals to the emotional system may work to get someone interested in an issue in the short term, but that it is hard to retain that level of interest. Unless they are given reasons to remain engaged, people's attention easily shifts to other issues.

3. Studies also show that the effects of worry can lead, paradoxically, to **emotional numbing**. This occurs after repeated exposures to emotionally draining situations and is a commonly observed reaction in individuals living in war zones or dealing with repeated hurricane threats in a short period. The danger of overexposure to threatening issues is especially high given the modern media environment where people confront a bewildering number and diversity of emotional experiences every day, ranging from news stories to sensational movies.[38]

HOW TO AVOID NUMBING AN AUDIENCE TO CLIMATE CHANGE

Climate change communicators should:

- Decide what portfolio of risks they want to make the public more aware of and then demonstrate the connection between those risks, such as the relationship between climate change and disease.
- As described in Section 3, balance information that triggers an emotional response with more analytic information to leave a mark in more than one place in the brain.
- Acknowledge that the audience has other pressing issues. Create a balance between pre-existing concerns and the climate change issues to be discussed.
- Gauge an audience's degree of numbing (i.e., ask them questions about their levels of media exposure to climate change, show them well-known images associated with climate change and note their reaction), make them aware of the various effects of numbing, and encourage them to briefly consider their level of worry and potential numbness to climate change.

WHAT IS THE SINGLE ACTION BIAS?

In response to uncertain and risky situations, humans have a tendency to focus and simplify their decision making. Individuals responding to a threat are likely to rely on one action, even when it provides only incremental protection or risk reduction and may not be the most effective option. People often take no further action, presumably because the first one succeeded in reducing their feeling of worry or vulnerability. This phenomenon is called the **single action bias**.[39]

For example, although recycling is important, it should be but one activity in a series of behavior changes aimed at reducing climate change. Switching to wind or other renewable energies, consuming less meat, conserving daily energy use, and eating locally grown food are other effective ways to mitigate climate change, to name but a few. However, if individuals and institutions participate in recycling programs, they may be prone to the single action bias and feel like they are already doing enough to protect the environment.

CRED research provides additional evidence of this phenomenon. One study found that, to adapt to climate variability, many farmers in Argentina engaged in only one activity to protect against the impact of drought on their livelihoods, despite having numerous options available to them. For instance, farmers who had the capacity to store grain on their farms were less likely to use irrigation or crop insurance although these measures would have added up to even greater protection against the impact of drought.[40]

Interestingly, recent polling may have found evidence of a mass single action bias—the election of President Barack Obama seems to have shifted Americans' attitudes about whether or not the state of the environment is improving. Nate Silver, of the polling blog FiveThirtyEight.com, argues that Democrats increasingly believe the environment is improving simply based on Obama's election, whereas the number of Republicans who say the environment is improving has remained about the same since 2008.

"Because of Barack Obama's election," Silver wrote, "many Americans assume that the environment is getting better, whether or not it actually is." Silver cited a Gallup poll from February 2009 that showed 41 percent of Americans think the environment is getting better, compared to just 26 percent in 2008. He argued that such perceptions could prove detrimental to legislative efforts to address global climate change and other environmental problems.[41]

HOW TO COUNTERACT THE SINGLE ACTION BIAS

It is human nature to fall prey to it and it is difficult to avoid, but there are steps that communicators can take to counteract the single action bias:

- Make an audience aware of the phenomenon. To demonstrate the single action bias, try the following exercise: Ask your audience how many of them have replaced their light bulbs with compact fluorescent lights—typically a large amount of people raise their hands.
- Then ask how many of them turn off their computer at night—again, a fair number of people will likely raise their hands. But if you ask who does both, the count will go down dramatically. Feel free to insert a third, fourth, or even fifth action to create a portfolio of energy saving and climate change mitigation behaviors.
- Provide energy-saving checklists that people can place in a prominent spot in their home or office. The checklists will remind and encourage people to go beyond just one tip. More people should take a diversified approach as a result.

The example below highlights a different but equally effective approach to defeating the single action bias.

EXAMPLE

The Simple Steps Campaign and Tiered Environmental Action

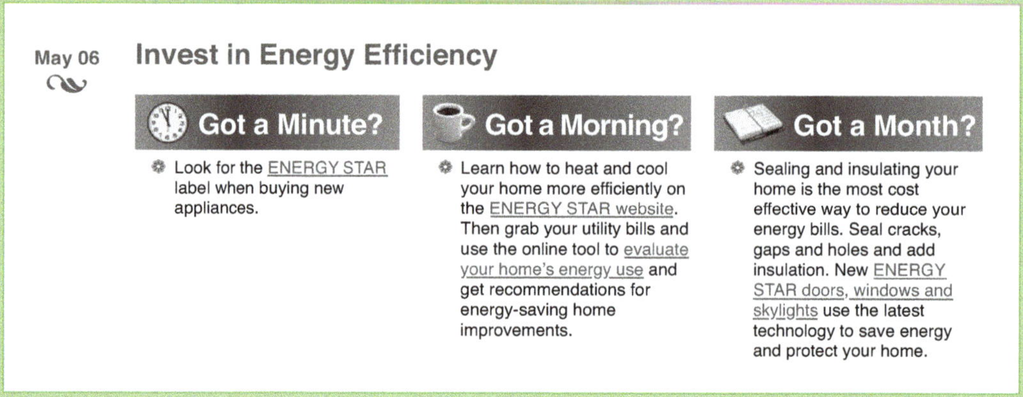

simplesteps.org

The Natural Resources Defense Council launched Simple Steps, a how-to campaign that divides environmental advice into three tiers based on the commitment level of its audience. Those interested in participating can select information based on whether they want to invest a minute, a morning, or a month adopting a more environmentally responsible lifestyle.

This campaign inherently acknowledges the influence of the single action bias and other psychological phenomenona that prevent people from taking effective action to solve complex problems. The campaign sought to counteract the single action bias by encouraging participants to increase their commitments incrementally.

5 Address Scientific and Climate Uncertainties

WHY ARE THERE UNCERTAINTIES IN CLIMATE SCIENCE?

As with other branches of science, climate science involves scientific uncertainty. Beyond that, however, uncertainty in climate science derives from the many complex forces that govern the earth's climate, from the axis of the planet's rotation to the changing composition of the atmosphere. Although scientists have gained significant insight into how the climate system functions, they do not have 100% confidence in their climate change projections—and they never will. What they can do, however, is make predictions based on the best available data, quantifying the uncertainties associated with those predictions.

Several areas of uncertainty exist in climate change prediction. One is due to the lack of complete knowledge of how the climate works, which will lessen with further study. Other uncertainty is due to natural variability in the climate system, which will not go away. And an additional element of uncertainty is due to the inability to predict human behavior and its cumulative impact on the earth's climate.

Future climate predictions depend on a number of changing variables in much the same way future traffic predictions do. Both systems operate under a certain level of volatility and uncertainty, but that does not prevent either climate scientists or traffic analysts from making forecasts with the information on hand. Although traffic forecasts days into the future may seem hard to trust, as are future climate projections for some people, both are determined by algorithms based on mass data from varying sources. A unique, location-specific model can provide greater accuracy for both traffic and climate scenarios. But with both systems, full certainty comes when it is already too late and one is in a jam.

THE PROBLEM WITH SCIENTIFIC UNCERTAINTY: THE HUMAN NEED FOR PREDICTABILITY

Because humans have a great need for predictability, uncertainty can be uncomfortable. Predictability helps people feel safe and secure, whereas uncertainty can lead to anxiety.[42] Predictability offers survival value. It provides control, helps people avoid threats to their physical and material well-being, and frees them from fear and anxiety. Furthermore, it allows people to plan and budget for the future. However, the human capability to prepare can be impaired by uncertainty.

Particularly when talking about complex topics like global climate change, it is important to find effective ways to communicate inherently uncertain information.

Too often discussions of climate science uncertainty convey the mistaken impression that scientists are hopelessly confused about this complicated subject, when in fact the uncertainties about exactly how much warmer the planet will be in 100 years do not change the very high confidence scientists have that human-made emissions of greenhouse gases are warming the planet and are likely to continue doing so.

To address this problem, IPCC scientists developed a "confidence terminology" to communicate estimates of uncertainty via everyday language. For example, "very high confidence" was used to refer to a prediction that has at least a nine out of ten chance of being correct. Other such terms included "high," "medium," "low," and "very low" confidence. "Very low confidence" referred to a prediction that had less than a one out of ten chance of being correct.

In cases where probabilistic estimates could be made, the IPCC also used "likelihood terminology" to define the likelihood of an outcome or result. "Virtually certain" had the highest likelihood with a greater than 99 percent probability, while "likely" denoted only a 66 percent or more probability of occurrence.[43]

Although such terms have greatly permeated public discourse on climate change, there is evidence that suggests people interpret such probability descriptors more subjectively than scientists intend.

For example, in a recent report's Summary for Policymakers, the IPCC stated, "Most of the observed increase in global average temperatures since the mid-20th century is very likely due to the observed increase in anthropogenic GHG [greenhouse gas] concentrations."[44] From the use of the term "very likely" in this sentence, the IPCC meant that there is a 90 percent or greater likelihood that emissions of greenhouse gases from human activities have caused most of the global average temperature increase since the mid-20th century.

But in a study by researchers at the University of Illinois at Urbana-Champaign, people assigned lower likelihood values to the IPCC's descriptors compared

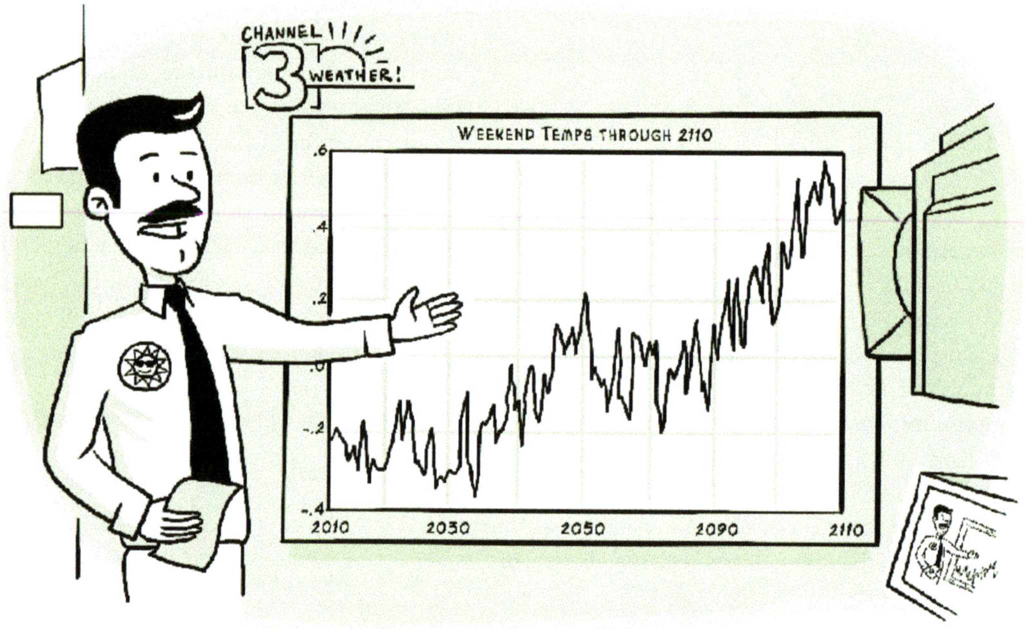

"So yes, Dan and Kathy, as you can see it looks like it'll be up and down until 2109, but you're certainly going to want to think about abandoning the planet after that...."

to what the IPCC actually meant. In other words, research subjects thought the scientific evidence of climate change was less conclusive, as conveyed by the IPCC's probability terms, than what the scientists had really reported. Among other recommendations, the researchers suggested that the IPCC consider including the associated range of probabilities whenever a probability descriptor is used, rather than only publishing a key to the terminology.[45]

HOW TO COMMUNICATE CLIMATE CHANGE UNCERTAINTY

Climate change uncertainties vary in type and significance, and they are difficult to convey without seeming to minimize the importance or understanding of the issue. One of the first key tasks for communicators is to put that uncertainty into context by helping audiences understand what is known with a high degree of confidence and what is relatively poorly understood.

In particular, scientists found that the general public interprets certain common words differently than do the scientists who used them.

THE NEED FOR PRECISION

Table 4 on page 27 shows a list of common words used to describe climate change that mean different things to scientists and the general public.[46]

Jargon filled explanations of uncertainty can easily undermine a scientist's message. For example, Senator Jim Inhofe asserted in a speech to the Senate that, "statements made by the National Academy of Sciences (NAS) cannot possibly be considered unequivocal affirmations that man-made global warming is a threat."[47] As evidence, he quoted the National Academy of Sciences 2001 report, dwelling on such phrases as "considerable uncertainty in current understanding," "estimates should be regarded as tentative and subject to future adjustments," "because of the large and still uncertain level of natural variability," "uncertainties in the time histories of various forcing agents," "cannot be unequivocally established."

Such phrases can easily translate as unreliable climate science to the greater public. Using the word *considerable* to describe uncertainty creates a disparity in meaning between common language and science. What quantity is "considerable"? This word is subject to varying interpretations. Similarly, the word *error* means *mistake* to most people, which is wholly different from the scientific definition of "error." Discussing uncertainty with unspecific language can lead to an unintentional overstatement and consequent criticisms.

Most critically, communicators should suggest neither more, nor less scientific certainty about climate change than actually exists. When significant uncertainty remains about a specific effect, they should explain why that uncertainty exists (e.g., the systems involved are so complex that science has yet to understand them sufficiently).

INVOKE THE "PRECAUTIONARY PRINCIPLE"

It is also important to recognize and emphasize that scientific uncertainty alone is not an adequate justification for inaction or business-as-usual policies and behaviors. Rather, it suggests that, at a minimum, it would be prudent to develop contingency plans and adopt adaptive management strategies. This would be in accordance with the "precautionary principle," which holds that action should be taken to reduce the risk of harm to the public from potential threats such as climate change, despite the absence of 100 percent scientific certainty about all aspects of the threat.

The precautionary principle has been considered internationally, including the 1992 United Nations Framework Convention on Climate Change, which states that countries should "take precautionary measures to anticipate, prevent or minimize the causes of climate change and mitigate its adverse effects. Where there are threats of serious or irreversible damage, lack of full scientific certainty should not be used as a reason for postponing such measures…"

Governor Arnold Schwarzenegger of California referred to the principle with a metaphor when he said: "If 98 doctors say my son is ill and needs medication

TABLE 4

Words with Different Meanings to Scientists and the General Public

SCIENTIFIC WORDS	NON-SCIENTIFIC MEANING	BETTER WORDS
Enhance	Improve	Intensify, increase
Uncertainty	Not knowing	Range
Risk	Low-probability event	Probability
Error	Wrong, incorrect	Uncertainty associated with a measuring device or model
Bias	Unfair and deliberate distortion	Offset from the observed value
Positive trend	A good trend	Upward trend
Positive feedback	Constructive criticism	Self-reinforcing cycle, vicious circle
Theory	A hunch, opinion, conjecture, speculation	Physical understanding of how this works
Hypothesis	Conjecture	Framework for physical understanding
Sign	Indication	Positive/negative value, plus/minus sign
Values	Ethics, money	Numbers, quantity
Manipulation	Exploitation	Changes in experimental or model conditions to study the impact of that condition
Scheme	Conspiracy	Blueprint
Productivity	Working hard	Photosynthesis
Anomaly	Abnormal occurrence	The deviation from a long term average

and two say 'No, he doesn't, he is fine,' I will go with the 98. It's common sense—the same with climate change. We go with the majority, the large majority....The key thing now is that since we know this industrial age has created it, let's get our act together and do everything we can to roll it back."[48] In this example, Schwarzenegger conveyed information about climate change risk and uncertainty in terms his audience could relate to.

The precautionary principle is a key consideration for making decisions under uncertainty, and it is useful to address potential harms that are outside of the environmental arena as well, as the following example illustrates.

THE BENEFITS OF TALKING ABOUT CLIMATE CHANGE INFORMATION IN GROUPS

Extensive anecdotal evidence from CRED's work with farmers in Africa and its laboratory studies suggest that people may understand probabilistic information better when it is presented to a group, where members have a chance to discuss it, rather than as individuals who have to try to understand it alone.[51]

Group processes allow individuals with a range of knowledge, skills, and personal experience to share diverse perspectives and work together on a problem.

EXAMPLE

Michigan Cherry Growers and Climate Uncertainty

Cherry blossoms have begun to appear seven to ten days earlier in Michigan than they did three decades ago, leaving them susceptible to potentially devastating spring frosts.[49] In 2002, a spring frost destroyed 99 percent of the crop, and cherry farmers wanted to know if these occurrences were likely to increase. They needed to make decisions about their $44 million-a-year-industry despite this climate uncertainty. And because a cherry tree can take up to a decade to bear fruit and typically has only a 20-30 year cycle of productivity, the farmers needed both extended and highly localized climate change information.

Simon Koopmann, commons.wikimedia.org

A group of agricultural experts, economists, climate scientists, and others began working to bring these cherry growers and other stakeholders information about climate change on a very local level. A single concrete climate prediction wasn't feasible. Instead these researchers needed to determine a wide range of climate scenarios for that region extending through the rest of the century. Further, they needed to communicate to the farmers their level of confidence per scenario. The farmers could then decide how to proceed, choosing to invest in wind machines or other frost protection, plant a hardier variety of cherry, switch to a different crop, or get out of farming altogether based on shifts in probability. Their livelihood depends on making sound decisions using the best available, yet still uncertain, scientific information.[50]

Group discussion provides a greater chance that multiple sources of information—both experiential and analytic—will be considered as part of the decision-making process. More energy is devoted to implementing solutions after group discussion. Furthermore, group context increases awareness of social support and activates social goals (see Section 6).

The example (below, left) illustrates how group discussion led to both better understanding of a probabilistic climate forecast and to generation (and eventual implementation) of more sound agricultural coping strategies.

As the example (below, right) shows, communicators should point out the probabilistic nature of climate science models and, when possible, engage and encourage group discussion about the uncertainties associated with climate change.

EXAMPLE

African Farmers and Climate Information

Carla Roncoli

Over the last decade, CRED researchers have been working with African farmers to interpret climate forecasts for use in agricultural decisions. In one study, farmers who attended climate discussion meetings had more ideas about potential adaptive responses to forecasts than those who did not attend. The study highlighted the importance of discussion as a way to understand and incorporate climate uncertainty into planning. The participatory process facilitated the understanding and use of climate information, allowing group members to pool their ideas and to plan coordinated responses. In several groups, the farmers commented that before they heard the forecast, they were uncertain about what course the seasonal rains would take and hence about what agricultural strategies to pursue. They remained unsure about what was coming and what to do as they heard different opinions voiced at the meetings, but once a consensus was reached, they trusted the forecast, and worked hard and effectively at the particular strategies the group had settled on.[52]

EXAMPLE

CRED Lab Experiment on Group Learning Processes

In order to study probabilistic learning CRED created a game in which students (either individually or in groups) learned about the probabilities for livestock to have a mosquito-borne illness called Rift Valley Fever (RVF). Students were then asked to play a game in which they bought and sold livestock that may or may not have RVF. In one option, which represented the optimal strategy, students could pay to test the animals before buying them. Students who learned about the probabilistic nature of the risk of RVF in a group were more likely to pay for the test, which maximized the joint outcome in the game, rather than try to achieve greater individual outcomes. The results strongly suggest that effective training requires both a cognitive and social component for people to recognize an optimal strategy.

The groups also showed a greater tendency to reframe information (from analytic to experiential and vice versa); provided additional opportunities to teach and learn from each other; and enabled the development of both social norms and shared goals.[53] Climate change communicators seeking to work with groups should set these as goals for their efforts as well.

6 Tap into Social Identities and Affiliations

The Tragedy of the Commons theory is as old as Aristotle, who said: **"That which is common to the greatest number has the least care bestowed upon it."**

WHAT IS THE TRAGEDY OF THE COMMONS?

The tragedy of the commons presents a conflict over resources between individual interests and the common good. Commons dilemmas describe conflicts resulting from free access and unrestricted demand for a finite natural resource. This ultimately threatens the resource and leads to exploitation. The benefits of exploitation go to individuals, each of whom is motivated to maximize his or her use of the resource, while the costs of exploitation are distributed among all who share the resource.[54] Overfishing of the world's fish populations and pollution of the earth's atmosphere are modern day examples of a "tragedy of unmanaged commons."

Environmental decisions pose a similar dilemma to the tragedy of the commons scenarios, in that an individual's benefit may or may not be the same as what benefits society. In other words, deciding to engage in behaviors that help mitigate climate change, a benefit for society, may seem more of a cost than a benefit to the individuals who would engage in them, at least in the short term. Climate change communicators need to recognize this dichotomy and address it by tapping into

multiple identities in their audiences, creating a sense of affiliation with each other, the environment, and the society that enjoys the benefits of its natural resources.

HOW TO TAP INTO GROUP IDENTITY TO CREATE A SENSE OF AFFILIATION AND INCREASE COOPERATION

An individual comprises numerous roles and identities, each of which has its own set of goals. In any given situation, an individual may call into play multiple identities (household member, town resident, CEO, parent, member of religious organization), even when the goals of the various identities may conflict with each other. To resolve that conflict, an individual has to decide which identity is most relevant in a situation.[55] The strength of affiliation that someone feels toward other members of a group (or the people that may be affected by a decision) can determine which identity that person chooses to apply in a particular situation.

When people make decisions, they recognize the situation, their identity in that situation, and the rules that are most appropriate given the situation and their chosen identity.[56] CRED research suggests that group affiliation may influence whether an individual decides to cooperate in a group decision or not for several reasons:[57]

- Group affiliation can activate social goals (i.e., concern for others, maximizing the good of the group);
- Participating in a group allows group norms to exert a stronger influence on individuals;
- Participating in a group also leads to greater intrinsic reward for individuals when group goals are achieved.

People who feel an affiliation with a group are thus more likely to cooperate in environmental decisions, such as joining a town's efforts to reduce greenhouse gas emissions. Further, people may continue such behaviors due to the "reward" found in helping the group reach its climate change goals, as demonstrated in the example on the right. Although any appeal to group identity

EXAMPLE

CRED Lab Experiment on Group Affiliation and Cooperation

CRED researchers designed an experiment to measure the effect of social goals, in particular the effect of affiliation on cooperation.[59] Students were split randomly into four-person groups (analogous to four large greenhouse gas emitters). The researchers created different levels of affiliation among the group members (temporary, short-lived connections). Groups then played a game that rewarded those who chose to defect rather than cooperate. CRED researchers found the following: that as affiliation increased, so did cooperation; that affiliation made social goals (e.g., the concern for others) a greater priority; and that the added benefit of cooperation more than made up for the sacrifice (in this case: monetary sacrifice). Students reported that they felt good about cooperating. Communicators who want to promote cooperation should try to activate social goals by integrating social and economic goals and by emphasizing an affiliation among group participants.

CRED research also suggests that local "messengers" (both individuals and institutions) may be more likely to get a response for calls to action on climate change than emissaries from distant locales. People are more likely to take action when they feel a strong sense of affiliation with the individual or institution making the request. Communicators from "out of town" may want to enlist someone locally known to help create a connection with their audience.

can help trigger group goals and cooperation, affiliations with smaller groups, such as a sorority or house of worship, can be stronger than those with larger groups, such as a political party or country.[58] Communicators will find it effective to create a sense of group affiliation within an audience, and they should try to find the most common yet smallest affiliation that the audience can identify with.

The example (below, left) illustrates the power of a local organization tapping into area residents' identity with the city to motivate new behaviors to help mitigate climate change. It also shows the importance of rewarding individual actions taken toward a group goal to reinforce such behaviors. The example (below, right) illustrates the power of tapping into social identities and creating "green" social norms.

EXAMPLE

Knoxville, Tennessee, Greens Up

Knoxville's "Make Downtown Green, Block by Block" campaign achieved great success by drawing on city identity. The Knoxville Utilities Board (KUB) and the city of Knoxville, along with their initiative partners, engaged downtown residents and businesses to purchase 400 blocks of green power, representing the 400 city blocks of downtown Knoxville. Through the Tennessee Valley Authority's Green Power Switch program, KUB now provides downtown Knoxville with energy created by renewable resources. The city celebrated in the spring of 2009 with comments from the mayor and recognition of the downtown residents and businesses that participated. KUB distributed 400 dogwood saplings during the event in honor of these environmentally-committed customers. This campaign emphasized people's identity with the city, utilized local messengers, and acknowledged the participating members of the community, providing a social incentive for others to act.[60]

EXAMPLE

The Energy Smackdown: Using Reality TV to Inspire Lower Energy Consumption

The Energy Smackdown, a reality television series, engages the greater public on the issue of climate change by showcasing what citizens of a community can do to reduce their own energy consumption. In season two, teams of households from three different communities in Massachusetts—Arlington, Cambridge, and Medford—competed to see which community could make the biggest energy reduction over 12 months. The "challenges" included biking to work, weatherizing their homes, eating locally grown food, and replacing shower fixtures and light bulbs with eco-friendly alternatives—all simple steps for the greater American audience to emulate. In addition, contestants were expected to talk to other community members about reducing carbon emissions. The first-place winners reduced their household consumption of energy by a whopping 73%. This contest simultaneously tapped into the contestants' identity within the household, the neighborhood, and the town and created new "green" social norms for all of the participating towns and possibly for viewers across America.

7 Encourage Group Participation

Sometimes climate change communicators need to go beyond presenting to a general audience to brokering an environmental decision within a group setting. Many environmental decisions are group decisions, so it is vitally important for communicators to understand how people participate in group settings, whether public or "closed door." Some of the variables include: the relationships that exist among the individuals and groups involved; the participants' individual and group goals; the different ways people participate in groups; and norms concerning how the meeting should be run.

EXAMPLE

Ugandan Farmers' Wives and Nonverbal Group Participation

Carla Roncoli

CRED research on farmer's decisions in the face of climate uncertainty in Uganda highlights that non-verbal behaviors during discussions are also forms of participation. Spatial arrangements reflect differences in social roles or power, which in turn affect how people participate. Ugandan men often sit close to the speakers, while the women sit on the margins of the group, tending children and other tasks. Although some women may directly address the group (particularly if called upon), they more often talk among themselves or communicate through non-verbal means, such as glances, clapping, or laughter. Gender and social position are important for how one participates and how others regard one's contributions.[61]

UNDERSTANDING THE MANY WAYS PEOPLE PARTICIPATE IN GROUPS

Norms about what happens in meetings are important because they determine who speaks when, how information is presented, and how people should disagree. Some people are more comfortable presenting from their experience, and this information should not be devalued because it is not "factual." There are also norms concerning language use: for instance, using technical language may seem rude when it makes the information being conveyed inaccessible to less-educated participants, essentially limiting their involvement in the discussion and, ultimately, the decision(s). There are similar norms concerning the meeting's end goal—in some cultural contexts, reaching group consensus may have a higher value than representing differences and allowing everyone to express their opinion.[62]

Eliciting participation from all of the various stakeholders is extremely important when trying to broker environmental decisions. Stakeholders who feel like they were part of the decision-making process are more likely to support the outcome. Early participation in the decision-making process is also a vital step in identifying the key problems that require solutions.

The example above indicates how understanding each audience member's particular form of participation can help communicators better judge if all members of the audience are taking part (in some form) in the discussion.

HOW TO SET THE STAGE FOR EFFECTIVE GROUP DISCUSSIONS OF CLIMATE CHANGE

Presentations on climate change are often filled with dense information that may leave group members with numerous questions and concerns. When organizing meetings with a diverse group of stakeholders, the most vital thing to remember is to allow ample time for discussion. Anecdotal evidence suggests that breaking large groups into smaller groups can help initiate discussion.

The example below shows the successful application of participatory processes to natural resource management in Florida. The box on page 36 provides tips for encouraging group participation.

EXAMPLE

Lobster Fishing in the Florida Keys

Hartmut Inerle, commons.wikimedia.org

The management of the spiny lobster fishery in the Florida Keys provides a great example of how to resolve a potential "tragedy of the commons" by encouraging the participation of all the various stakeholders. Social scientists Michael Orbach and Jeffrey Johnson worked with the commercial fishing industry, recreational fishermen, environmentalists, and others to solve the problem of having too many traps in the water.

With the input of fishermen and other stakeholder representatives, they gathered a wide range of opinions about the issue through participatory observation, mail surveys, and personal interviews. The researchers then held three series of workshops, each drawing up to 200 stakeholders and other interested members of the community.

The first series of workshops discussed whether there was a problem in the fishery and presented general information about limited entry systems (i.e., how many traps can be placed in the water), with specific examples. In the second series of workshops, participants developed a ranking system that they then applied to several possible alternatives to solve the problem according to specific criteria, which they also developed. In the third series the participants summed up the relative effects of their alternatives. The outcome of the third series of workshops was a recommendation to go with a transferable trap certificate program.

The program, which required Florida to implement new legislation, was actually lobbied for by the stakeholders—commercial and recreational fishermen and environmentalists! Through a participatory process, the group reached its main goals: to reduce the number of traps by over half while keeping the catch relatively constant. This successful shift increased the overall profitability of the fishery, reduced conflict, and made the fishery easier to manage.[63]

Ways To Encourage Group Participation

- Know your audience. Be aware of differences among participants (i.e., education, wealth, occupation, ethnicity, religious motivations, understanding of participation norms). Consider who has authority and who does not.

- Clarify expectations for the meeting and the role of the group in the decision(s) to be made. Will the group make the final decision, or are members only advising a decision-making body elsewhere? How will the final decision be made?

- Try to involve as many people, or representatives of as many groups, as possible. Be aware of how people were invited to attend and if anyone was overlooked.

- Think about how the group processes information. Will the individuals meet for the first time in a large group setting or will they meet ahead of time on a more informal basis to discuss issues? If so, communicators may want to distribute information ahead of time to give them time to review and prepare for the formal meeting.

- Recognize different forms of participating. Work to include all voices—give everyone a chance to speak, and respect different methods for making an argument.

- Make sure all viewpoints are represented. Solicit ideas from the different individuals involved in the discussion.

- Utilize pre-existing relationships within the group to communicate information since information passed through familiar channels is often most effective.

- Be conscious of verbal and non-verbal means of participation. Appreciate non-linguistic means of communication such as disruptive behavior, nods of approval, or applause.

- Acknowledge that participants will have other goals. Meetings are often a place for people to socialize, meet new people, and advance their own personal or political goals. These "social goals" are a necessary part of participation.

- Be aware that members of a group will interact outside of meetings. A meeting is only one part of the whole decision process.

- Use group discussion to generate solutions. People may be more willing to acknowledge a problem if they feel there are solutions to dealing with the problem. This can help keep messages positive, encourage optimism, and demonstrate how groups can be a powerful force in tackling the climate change challenges ahead.

8 Make Behavior Change Easier

Climate change communicators often end their presentations encouraging audiences to make changes in their behavior that will help mitigate the effects of climate change. This section addresses how policymakers, business leaders, and environmental organizations can make such behavior changes easier by taking advantage of **default effects** (the human tendency to stick with the option that is selected automatically instead of choosing an alternate option), including making environmentally responsible behaviors the default option as often as possible.

UNDERSTANDING DEFAULT EFFECTS ON DECISION MAKING

It is important to consider default effects when people make decisions over time. For example, when people have a choice between Option A, with benefits and costs in the present, and Option B, whose benefits and/or costs might not be realized until some point in the future, the default option can affect their preferences. Particularly when making decisions about consumption (to purchase something, to receive a reward, to make a sacrifice), people tend to be more patient when the default option is to wait versus when the default option is to receive something now.[64]

HOW TO OPTIMIZE THE DEFAULT OPTION

When presenting a choice with multiple options, it is important to pay attention to the default option. If option A is the default and a person wants A, it is already chosen. But if a person wants B, he or she will have to make an effort to switch from A to B. Because the default option requires no action, it is always easier, and so people tend to accept it whether or not they would have chosen it if it were not the default option. By making socially beneficial choices the default option, policymakers can positively influence individual decisions concerning natural resources like air or water.[65] Page 38 shows an example of this principle in action.

One German study showed that changing defaults could promote green sources of energy. The study also found that the way information is presented, specifically for the default option, can strongly affect people's choice of electricity, and that they tend to use the kind of electricity that is offered to them as the default. In the first laboratory experiment, more participants chose the green utility when it was the default than when the "grey" utility was the default. In the second laboratory experiment, participants displayed an attachment to their default, asking for more money to give up green electricity than the amount they would have paid for it.[66]

EXAMPLE

How Rutgers University Saved 1,280 Trees in One Academic Year

Erich Nagler

After deciding that the university computer labs were wasting too much paper, Rutgers University simply made double-sided printing the default option on its lab printers. This tiny act saved 7,391,065 sheets of paper in the first semester, or roughly 620 trees for the semester, and 1,280 trees for the academic year. Students, who frequently have no preference, must now manually select the option to print on only one side of the page. The option to conserve is made that much easier by becoming the default option.[67]

PROVIDE NEAR-TERM INCENTIVES

Giving people an immediate incentive, if possible, also makes behavior change easier. For instance, the prospect of saving money over the next 20 years by weatherizing one's home may make economic sense, but may not effectively motivate action. In contrast, giving an immediate incentive can serve as an effective driver. For example, when presenting to a church, school, or community center group, climate change communicators can publicize the names of those who sign up for weatherization, thus providing an immediate social incentive to supplement the delayed economic incentive.

By using an economic incentive, the Japanese government significantly increased the demand for green vehicles. The government provided "scrap incentives," either tax breaks or rebates, for consumers to scrap their old cars and buy ecological vehicles. The average consumer may discount the long-term savings of driving a hybrid, but will readily appreciate such immediately tangible (in this case, financial) benefits.[68]

Conclusion

Gaining public support for climate change policies and encouraging environmentally responsible behavior depends on a clear understanding of how people process information and make decisions. Social science research provides an essential part of this puzzle but, as the guide makes clear, there is no "one-size-fits-all" approach to the challenges of communicating about climate change. Rather, each of the many barriers presents a new opportunity to improve the way we present information about climate change and the behaviors required to mitigate it.

It is our hope that readers will use the information in this guide—paired with the latest advances in climate science, engineering, economics, and environmental policy—to communicate climate change in a way that resonates with their audience. Ensuring that people feel both a personal connection with climate change and a desire to take action to mitigate its impact, without becoming overwhelmed by the scale of the problem, is key. Whereas it goes beyond the expertise of the authors of this social-science–based guide to provide specific policy recommendations and other climate change solutions, climate change policies are an essential component about which the public needs to be informed and for which political support needs to be generated. With an issue as complex as climate change, people need to know there are solutions to dealing with it, and that they can be part of those solutions.

Social science research provides compelling evidence for an optimist's view that climate change communicators can reach both policymakers and the public, informing and inspiring them to address climate change.

Following is a brief summary of the principles discussed in detail in the guide. We encourage readers to use the summary as a reference and to refer back to the guide for more in-depth information about topics that interest them.

The Principles of Climate Change Communication in Brief

❶ KNOW YOUR AUDIENCE

- **Mental models** represent a person's thought process for how something works. They help shape risk perceptions, actions, and behavior; influence what people pay attention to in complicated situations; and define how people approach and solve problems. Mental models serve as the framework into which people fit new information.
- A **confirmation bias** makes people look for information that is consistent with what they already think, want, or feel, leading them to avoid, dismiss, or forget information that will require them to change their minds and their behavior.
- People often exhibit a strong preference for their existing mental models about climate change, making them susceptible to confirmation biases that lead them to misinterpret or even refute scientific data.
- Mental models are not static—people can update them by correcting misinformation, inserting new building blocks, and/or making new connections with existing knowledge.

Tip: • *Discover what misconceptions the audience may have in their mental models about climate change. "Disconnect" the erroneous climate change information from other parts of the model and replace it with new facts.*

❷ GET YOUR AUDIENCE'S ATTENTION

- **Framing** is setting an issue within an appropriate context to achieve a desired interpretation or perspective.
- Framing is not intended to deceive or manipulate people, but to make credible climate change information more accessible to the public.
- Framing can be a subtle art—even the choice of a single word can make the difference between winning and alienating an audience.
- People feel better and more positive about achieving their goals and are more likely to sustain their behavior when their goals are framed in a manner that feels naturally comfortable to them.
- People with a **promotion focus** see a goal as an ideal and are concerned with advancement. They prefer to maximize or increase gains.
- People with a **prevention focus** see a goal as something they ought to do and are concerned with maintaining the status quo. They prefer to minimize or decrease losses.
- People tend to discount the importance of future events. Many people count environmental and financial consequences as less important with every year they are delayed.
- People have a natural tendency to avoid losses rather than to seek gains. They tend to discount future gains more than future losses.

Tips:
- Consider the audience's membership in specific subcultures (groups of people with distinct sets of beliefs, or based on race, ethnicity, class, age, gender, religion, occupation, etc.).
- Select a frame/frames that will resonate with your audience.
- Prepare numerous frames ahead of time (i.e., climate change as a religious, youth, or economic issue).
- When framing climate change, be careful not to focus so intently on one particular aspect that the audience loses sight of the bigger picture.
- Consider people's goals when framing a message. Tailoring messages to people's natural promotion and prevention orientations increases the level of response for both groups.
- See Words That Appeal to Those With Either a Promotion or Prevention Focus on page 8 and include both types of wording when crafting messages.
- Bring the message close to home. Highlight the current and potential impacts of climate change not only globally, but also locally to increase the audience's sense of connection with the issue.
- Leverage local extreme weather events, using them as "teachable moments" during which to relate climate change to the experience of your audience. (However, keep in mind that although climate change may increase the chance that a particular event will occur, it does not cause an event to take place.)
- Tap into people's desire to avoid future losses rather than realize future gains.
- Present information in a way that makes the audience aware of potential current and future losses related to inaction on climate change instead of focusing on current and future gains.
- Remember that audiences may be more likely to make changes to their behavior if climate change information is framed as "losing a little bit now instead of losing much more in the future."

3 TRANSLATE SCIENTIFIC DATA INTO CONCRETE EXPERIENCE

- Attempts to convey the immediacy of the climate change challenge have fallen short of translating climate change into a near-term (as well as a long-term) danger on par with other imminent societal and personal threats.
- Many of the highly publicized graphs and charts showing global climate change data fail to inspire a sense of urgency in many audiences.
- Psychologically, distant risks do not set off the same alarms that immediate risks do. Human minds are not designed to immediately react to threats like climate change that seem to manifest themselves in the distant future.
- The human brain has two different processing systems: the experiential processing system, which controls survival behavior and is the source of emotions and instincts, and the analytical processing system, which controls analysis of scientific information (see Table 2 on page 16).
- Despite evidence that the experiential processing system is the stronger motivator for action, most climate change communication remains geared towards the analytical processing system. Personal or anecdotal accounts of negative climate change experiences, which could easily outweigh statistical evidence, are rarely put into play.
- Low comprehension of or interest in communications laden with scientific language may also contribute to the public's lack of response to climate change messages.

Tips:
- When creating presentations on climate change, use experiential tools such as:
  - Vivid imagery, in the form of film footage, metaphors, personal accounts, real-world analogies, and concrete comparisons and
  - Messages designed to create, recall, and highlight relevant personal experience and to elicit an emotional response.
- A message that combines elements that appeal to both the analytic and experiential processing systems will best reach and resonate with an audience.
- Avoid using jargon, complicated scientific terms, and acronyms when talking to the general public. Instead, use words that will make sense to the audience (see Examples of Simplified Scientific Terms on page 19).
- Sometimes only a scientific term is sufficient for getting a point across. In that case, thoroughly define the term for the audience. Remember that stringing together too many scientific terms and acronyms, even if well-defined, may cause the audience to spend their time and mental energy deciphering vocabulary instead of absorbing the overall point.

4 BEWARE THE OVERUSE OF EMOTIONAL APPEALS

- Although an emotional appeal may increase an audience's interest in a climate change presentation in the short run, it may backfire down the road, producing negative consequences that often prove quite difficult to reverse.
- The **finite pool of worry** refers to the limited capacity people have for worrying. As worry increases about one type of risk, concern about other risks may lessen. People have a limited capacity for how many issues they can worry about at once.
- Appeals to the emotional system may work in the short term, but it is hard for people to retain that level of emotional intensity. People's attention can easily shift to other issues unless they are given additional reasons to remain engaged.
- The effects of worry can lead to **emotional numbing**, which occurs after repeated exposures to an emotionally draining situation.
- Individuals reacting to a threat are likely to employ only one response, even when it provides only incremental protection or risk reduction and may not be the most effective option. People often take no further action, presumably because their first response succeeded in reducing their feeling of worry or vulnerability. This is called the **single action bias**.

Tips:
- *See How to Avoid Numbing an Audience to Climate Change on page 21.*
- *See How To Counteract the Single Action Bias on page 23.*

5 ADDRESS SCIENTIFIC AND CLIMATE UNCERTAINTIES

- Although scientists have gained significant insight into how the climate system functions, they do not have 100% confidence in their climate change projections—and they never will. What they can do is make predictions based on the best available data, quantifying the uncertainties associated with those predictions.
- Because humans have a great need for predictability, uncertainty can be uncomfortable.
- Climate science uncertainty often conveys the mistaken impression that scientists are hopelessly confused about this complicated subject, when in fact scientific uncertainties about exactly how much warmer the planet will be in 100 years does not change the very high confidence scientists have that human-made greenhouse gas emissions are warming the planet and are likely to continue doing so.
- Climate change uncertainties vary in type and significance and are difficult to convey without seeming to minimize the importance or understanding of the issue.
- People may understand probabilistic information better when it is presented to a group, where members have a chance to discuss it, rather than as individuals who have to try to understand it alone.
- Group processes allow individuals with a range of knowledge, skills, and personal experience to share diverse perspectives and work together to solve a problem.
- Group discussion provides a greater chance that multiple sources of information—both experiential and analytic—will be considered as part of a climate-change related decision-making process.
- People devote more energy to implementing solutions after participating in a group discussion.
- Group context increases awareness of social support and activates social goals (see Section 6 for more about the dynamics of group information processing and decision making).

Tips:
- *Put uncertainty into context and help an audience understand what scientists know with a high degree of confidence and what they have a relatively poor understanding of.*
  - *Overstated uncertainty or poorly worded explanations of uncertainty can easily undermine a message.*
  - *Suggesting either more or less scientific certainty than actually exists can confuse an audience.*
- *See Words with Different Meanings to Scientists and the General Public on page 27 to ensure your words are precise and convey what you intended.*
- *Invoke the precautionary principle by addressing the potential harms of climate change that lack full scientific certainty.*
- *Whenever possible, present climate change information to informal groups where people are free to ask questions and discuss issues with the speaker and each other.*

6 TAP INTO SOCIAL IDENTITIES AND AFFILIATIONS

- **Commons dilemmas** describe conflicts resulting from free access to and unrestricted demand for a finite natural resource. This ultimately threatens the resource and leads to exploitation. The benefits of exploitation go to individuals, each of whom is motivated to maximize his or her use of the resource, while the costs of exploitation are distributed among all who share the resource.
- In environmental decisions, an individual's benefit may or may not be the same as what benefits society.
- In any given situation, an individual may call into play multiple identities (parent, CEO, etc.), even when the goals of the various identities may conflict with each other. To resolve that conflict, an individual has to decide which identity is most relevant in a situation.
- The strength of affiliation that someone feels toward other members of a group can determine which identity that person chooses to apply in a particular situation.
- Affiliations with smaller groups can be stronger than those with larger groups.
- Local messengers may get a stronger response to calls for action on climate change than emissaries from more distant locales. People are more likely to take action when they feel a sense of affiliation with the individual or institution making the request.

Tips:
- *Tap into the multiple identities represented by your audience; bolster audience members' sense of affiliation with each other, the environment, and the society that enjoys the benefits of its natural resources.*
- *If communicating as an "outsider," enlist the aid of someone locally known to introduce you.*

7 ENCOURAGE GROUP PARTICIPATION

- Many environmental decisions are group decisions, so it is important for communicators to understand how people participate in group settings.
- Norms about what happens in meetings are important because they determine who speaks when, how information is presented, and how people should disagree.

Tips:
- *Eliciting participation from various stakeholders is important when trying to broker environmental decisions. Stakeholders who feel like they were part of the decision-making process are more likely to support the outcome.*
- *Encourage early participation in the decision-making process to ensure the group identifies the key problems that require solutions.*
- *Presentations on climate change are often filled with dense information that may leave audience members with numerous questions and concerns. When organizing meetings with a diverse group of stakeholders, leave ample time for discussion.*
- *Breaking large groups into smaller groups can help initiate discussion.*
- *See Ways To Encourage Group Participation on page 36.*

8 MAKE BEHAVIOR CHANGE EASIER

- Taking advantage of **default effects** (the human tendency to stick with the option that is selected automatically instead of choosing an alternate option) can encourage audiences to make changes in their behavior that will help mitigate the effects of climate change.
- When making decisions about consumption, people tend to be more patient when the default option is to wait vs. when the default option is to receive something now.
- Because the default option requires no action, it is always easier, and so people tend to accept it whether or not they would have chosen it if it were not the default option.

Tips:
- *By making socially beneficial choices the default option, policymakers can positively influence individual decisions concerning natural resources.*
- *Giving people an immediate incentive, if possible, makes behavior change easier.*

FURTHER READINGS

AAAS's Communicating Science: Tools for Scientists and Engineers

The American Association for the Advancement of Science's Center for Public Engagement provides resources for researchers wishing to improve communication with the greater public, offering online webinars, how-to tips for media interviews, and strategies for identifying public outreach opportunities, in addition to in-person workshops.

http://communicatingscience.aaas.org/Pages/newmain.aspx

Communicating on Climate Change: An Essential Resource for Journalists, Scientists, and Educators

This resource guide for editors, reporters, scientists, and academics, compiled by Bud Ward, is based on Metcalf Institute workshops dealing with communication between journalists and climate scientists. It provides tips and tools for covering climate change.

http://metcalfinstitute.org/Communicating_ClimateChange.htm

Creating a Climate for Change: Communicating Climate Change and Facilitating Social Change

With contributors from diverse professional backgrounds, this book looks at communication and social change specifically targeted to climate change. It provides practical suggestions on how to communicate climate change and how to approach related social change more effectively. This volume is of interest to academic researchers and professionals in climate change, environmental policy, science communication, psychology, sociology, and geography.

Moser, S. and Dilling, L., eds. (2007). *Creating a Climate for Change: Communicating Climate Change and Facilitating Social Change*, Cambridge: Cambridge University Press.

Futerra's Communications Tactics for Climate Change

This communications agency, working on corporate responsibility and sustainability, offers easy to understand communications techniques to prompt behavior change affecting climate change.

http://www.futerra.co.uk/downloads/NewRules:NewGame.pdf

Global Warming's "Six Americas"

A national study by the Yale Project on Climate Change and the George Mason University Center for Climate Change Communication identifies six distinct climate change groups within the American public, ranging from "the alarmed" to "the dismissive." This report profiles these six different audiences and suggests ways to improve education and communication efforts to engage them.

http://www.climatechangecommunication.org/images/files/Global_Warming's_Six_Americas_2009r.pdf

ICLEI's Outreach and Communications Guide

This online guide is designed to help local governments effectively communicate climate information to their constituencies. It contains an array of steps and methodologies for communication and outreach efforts, as well as a compilation of best practices from around the United States.

http://www.icleiusa.org/action-center/engaging-your-community/outreach-and-communications-guide

Making Climate Hot: Communicating the Urgency and Challenge of Global Climate Change

The article explains how to increase public understanding of, and civic engagement with, climate change, providing context for obstacles and seven strategies that applied together can increase wider public concern and build momentum for social and policy change.

Moser, S., Dilling, L. (2004). *Making the Climate Hot: Communicating the Urgency and Challenge of Global Climate Change*. Environment, Volume 26, Number 10, pp.32–46.

Nudge: Improving Decisions about Health, Wealth, and Happiness

This book, applicable to individuals and governments alike, describes how choice architecture, based on the understanding of how people think, can nudge us to make better choices about better health, sounder investments, and cleaner environments without limiting freedom of choice.

Thaler, R. H. and Sunstein C.R. (2008). Nudge: Improving Decisions about Health, Wealth, and Happiness, Yale University Press.

Psychology and Global Climate Change: Addressing a Multi-faceted Phenomenon and Set of Challenges: A Report by the American Psychological Association's Task Force on the Interface Between Psychology and Global Climate Change

For this report, APA's task force examined decades of psychological research and practice that have been specifically applied and tested in the arena of climate change. The report offers a detailed look at the connection between psychology and global climate change and makes policy recommendations for psychological science.

http://www.apa.org/releases/climate-change.pdf

The Scientist's Guide to Talking with the Media

This book teaches researchers how to deliver an accurate message to a broader audience through the media, providing tips on how to turn abstract concepts into concrete metaphors, form sound bites, prepare for interviews, and even become a reporter's go-to scientist.

Hayes, R. & Grossman, D. (2006). The Scientist's Guide to Talking with the Media: Practical Advice from the Union of Concerned Scientists. Rutgers: Rutgers University Press.

For a complete list of CRED publications, visit *cred.columbia.edu/decisionpolicymakers/publications/*.

REFERENCES

1 Leiserowitz, A. (2007a) American opinions on global warming. *A Yale University/Gallup/ClearVision Institute Poll.* New Haven, CT: Yale School of Forestry & Environmental Studies.

2 Economy, jobs, trump all other policy priorities in 2009. Pew Research Center for the People and the Press. (2009, January 22). Retrieved from http://people-press.org/report/485/economy-top-policy-priority

3 Carey, S. (1986). Cognitive science and science education. *American Psychologist*, 41(10), 1123-1130.

4 Morgan, M., Fischhoff, B., Bostrom, A., et al. (2002). *Risk Communication: A Mental Models Approach.* Cambridge: Cambridge University Press.

5 Will, G. (2009, February 15). Dark Green. *The Washington Post.* Retrieved from http://www.washingtonpost.com/wp-dyn/content/article/2009/02/13/AR2009021302514.html

6 Revkin, A. (2008, March 2). Skeptics on human climate impact seize on cold spell. *The New York Times.* Retrieved from http://www.nytimes.com/2008/03/02/science/02cold.html?emc=eta1#

7 Lorenzoni, I., Leiserowitz, A., De Franca Doria, M., et al. (2006, April). Cross-national comparisons of image associations with "global warming" and "climate change" among laypeople in the United States of America and Great Britain. *Journal of Risk Research*, 9(3), 265-281.

8 Leiserowitz, A. (2007b) Communicating the risks of global warming: American risk perceptions, affective images and interpretive communities. In S. Moser and L. Dilling, (Eds.) *Creating a climate for change: Communicating climate change and facilitating social change* (pp. 44-63). Cambridge: Cambridge University Press.

9 Scheufele, D.A., Nisbet, M.C. (2007). Framing. In L. L. Kaid & C. Holz-Bacha (Eds.), *Encyclopedia of Political Communication*. Thousand Oaks, CA: Sage.

10 Cash, D., Clark W., Alcock, F., et al. (November 2002). Salience, Credibility, Legitimacy and Boundaries: Linking Research, Assessment and Decision Making. *John F. Kennedy School of Government, Harvard University, Faculty Research Working Papers Series*, RWP02-046.

11 Broder, J. M. (2009, March 6). House bill for a carbon

tax to cut emissions faces a steep climb. *The New York Times*. Retrieved from http://www.nytimes.com/2009/03/07/us/politics/07carbon.html?_r=1&scp=1&sq=carbon%20tax&st=cse
Policy Options for Reducing CO_2 Emissions. *Congressional Budget Office*. (February 2008). Retrieved from http://www.cbo.gov/ftpdocs/89xx/doc8934/toc.htm
12 Hardisty, D. J., Johnson, E. J., Weber, E. U. (in press). A Dirty Word or a Dirty World? Attribute Framing, Political Affiliation, and Query Theory. *Psychological Science*.
13 Cesario, J., Grant, H., Higgins, E.T. (2004). Regulatory fit and persuasion: Transfer from "feeling right." *Journal of Personality and Social Psychology*, 86, 388-404.
Higgins, E.T. (1997). Beyond pleasure and pain. *American Psychologist*, 52, 1280-1300.
Higgins, E.T. (2000). Making a good decision: Value from fit. *American Psychologist*, 55, 1217-1230.
14 Appelt, K.C., Higgins, E.T. (2009). A regulatory focus coding scheme. Manuscript in preparation.
15 Leiserowitz (2007a), op. cit.
16 Weather and Climate Extremes in a Changing Climate. *Climate Science*. (2008). Retrieved from http://www.climatescience.gov/Library/sap/sap3-3/final-report/
17 Bryant, N. (2007, December 7). Australian Climate Policy in from the Cold. *BBC News*. Retrieved from http://news.bbc.co.uk/2/hi/asia-pacific/7126249.stm
18 Climate Change and Hurricane Katrina. Center for Media and Democracy. Retrieved July 29, 2009, from http://www.sourcewatch.org/
19 Broncaccio, D. (Narrator). (2009, April 19). On Thin Ice. *Now on PBS*. Retrieved from www.pbs.org/now/shows/516/index.html
20 Slovic, P. (2000). *The perception of risk*. London: Earthscan.
Slovic, P., Finucane, M. L., Peters, E., et al. (2004). Risk as analysis and risk as feelings: Some thoughts about affect, reason, risk and rationality. *Risk Analysis*, 24, 311-322.
Weber, E. U. (2006). Experience-based and description-based perceptions of long-term risk: why global warming does not scare us (yet). *Climatic Change*, 77(1-2), 103-120.
21 Hardisty, D. J., Weber, E. U. (2009). Discounting Future Green: Money Versus the Environment. *Journal of Experimental Psychology: General*. 138, 329-340.
22 Kahneman, D. & Tversky, A. (1979). Prospect theory: An analysis of decision under risk. *Econometrica* 47, 263-291.
23 Thaler, R. (1981). Some empirical evidence on dynamic inconsistency. *Economics Letters*, 8, 201-207.
24 Gore, A. (2007, December 10). Nobel Lecture. Nobel Peace Prize, Oslo, Norway. Retrieved from http://nobelprize.org/nobel_prizes/peace/laureates/2007/gore-lecture_en.html
25 Security Council Holds First-Ever Debate on Impact of Climate Change on Peace, Security, Hearing Over 50 Speakers. United Nations (2007, April 17). Retrieved from http://www.un.org/News/Press/docs/2007/sc9000.doc.htm
26 National security and the threat of climate change. The CAN Corporation. (2007). Retrieved from http://www.cna.org/nationalsecurity/climate/report/National%20Security%20and%20the%20Threat%20of%20Climate%20Change.pdf
Climate Change 2007: Synthesis report. IPCC. (2007). Retrieved from http://www.ipcc.ch/pdf/assessment-report/ar4/syr/ar4_syr.pdf
27 Indoor & Outdoor Air Pollution. Centers for Disease Control and Prevention. Retrieved from http://www.cdc.gov/nceh/globalhealth/projects/airpollution.htm
28 *Climate Change 2007: Synthesis Report: Contribution of Working Groups I, II and III to the Fourth Assessment Report of the Intergovernmental Panel on Climate Change Core Writing Team*. Pachauri, R.K. and Reisinger, A. (Eds.) IPCC, Geneva, Switzerland.
29 Pew Research Center for the People and the Press, 2009.
30 Chaiken, S., Trope, Y. (1999). *Dual Process Theories in Social Psychology*. New York: Guilford Publications.
Epstein, S. (1994). Integration of the cognitive and the psychodynamic unconscious. *American Psychologist*, 49, 709–724.
Marx, S.M. et al. (2007) Communication and mental processes: Experiential and analytic processing of uncertain climate information. *Global Environmental Change*, 17(1), 47-58.
Sloman, S.A. (1996). The empirical case for two systems of reasoning. *Psychological Bulletin*, 1 (119), 3–22.
Weber (2006), op cit.
31 Chaiken op cit.
Epstein, op cit.

Marx, op cit.
Sloman, op cit.
32 Loewenstein, G., Weber, E.U., Hsee, C.K. (2001). Risk as feelings. *Psychological Bulletin*, 127 (2), 267–286.
Slovic, P., Finucane, M., Peters, E., et al. (2002). The affect heuristic. In: Gilovich, D.G.T., Kahneman, D. (Eds.), *Intuitive Judgment: Heuristics and Biases*. Cambridge University Press, New York.
33 Marx, S., Shome D., Weber, E.U. (2006). Analytic vs. Experiential Processing Exemplified through Glacial Retreat Education Module. Center for Research on Environmental Decisions.
34 Linville, P.W., Fischer, G.W. (1991). Preferences for separating and combining events: a social application of prospect theory and the mental accounting model. *Journal of Personality and Social Psychology*, 60, 5–23.
35 Weber (2006), op cit.
36 Pew Research Center for the People and the Press, op cit.
37 Hansen, J., Marx, S., Weber, E. U. (2004). *The Role of Climate Perceptions, Expectations, and Forecasts in Farmer Decision Making: The Argentine Pampas and South Florida*. International Research Institute for Climate Prediction (IRI), Palisades, NY: Technical Report 04-01.
38 Linville, op cit.
Weber (2006), op cit.
39 Weber, E. U. (1997). Perception and expectation of climate change: Precondition for economic and technological adaptation. In M. Bazerman, D. Messick, A. Tenbrunsel & K. Wade-Benzoni (Eds.), *Psychological and Ethical Perspectives to Environmental and Ethical Issues in Management* (pp. 314-341). San Francisco: Jossey-Bass.
40 Hansen, op cit.
41 Silver, N. (2009, April 22). When Hope is the Enemy of Change. *FiveThirtyEight*. Retrieved from http://www.fivethirtyeight.com/2009/04/when-hope-is-enemy-of-change.html
42 Maslow, A.H. (1943). A theory of human motivation. *Psychological Review* 50: 370-396.
43 Solomon, S., Qin, D., Manning, M., et al. (2007). Technical Summary. In: *Climate Change 2007: The Physical Science Basis. Contribution of Working Group I to the Fourth Assessment Report of the Intergovernmental Panel on Climate Change*. Retrieved from http://www.ipcc.ch/pdf/assessment-report/ar4/wg1/ar4-wg1-ts.pdf
44 IPCC, 2007: Summary for policymakers. In *Climate Change 2007: The Physical Science Basis. Contribution of Working Group I to the Fourth Assessment Report of the Intergovernmental Panel on Climate Change*. Retrieved from http://www.ipcc.ch/pdf/assessment-report/ar4/wg1/ar4-wg1-spm.pdf
45 Budescu D.V., Broomell S., Por, H. H. (2009, March). Improving communication of uncertainty in the reports of the Intergovernmental Panel on Climate Change. *Psychological Science*, 20 (3), 299-308.
46 Hassol, S. J. (2008 March 11). Improving how scientists communicate about climate change. *Eos*, 89(11), 106-107.
47 Inofe J. (2005, April 8). First Four Pillars. Senate Floor Statement. Retrieved from http://inhofe.senate.gov/pressreleases/pillar.htm
48 Friedman, T. (2007, April 15). The power of green. *The New York Times*. Retrieved from http://www.nytimes.com/2007/04/15/magazine/15green.t.html?scp=1&sq=If%2098%20doctors%20say%20my%20son%20is%20ill&st=cse
49 McGlashen, A. (2009, May 6). Cherry growers deciphering climate models, find uncertainty reigns. *The Daily Climate*. Retrieved from http://wwwp.dailyclimate.org/tdc-newsroom/2009/05/Pushing-to-localize-global-climate-predictions
50 McGlashen, op cit.
51 Krantz, D.H., Peterson, N., Osgood, D.E., et al., (2006) Social interaction and risk sharing in probability learning (working paper).
Marx, op cit.
Patt, A., Suarez, P. and Gwata, C. (2005). Effects of seasonal climate forecasts and participatory workshops among subsistence farmers in Zimbabwe. *PNAS*, 102(35), 12623-12628.
Roncoli, C. (2006). Advances in ethnographic and participatory approaches to research on farmers responses to climate predictions. *Climate Research*, 33, 81-99.
52 Roncoli, C., Jost, C., Kirshen, P., et al., From accessing to assessing forecasts: an end-to-end study of participatory climate forecast dissemination in Burkina Faso (West Africa) *Climatic Change*, (forthcoming).
Roncoli, C., Orlove, B.S., Kabugo, M., et al. "Multiple Styles of Participation in Farmers' Discussions of Climate

Information in Uganda," *Agriculture and Human Values*, (under review).
53 Krantz (2007), op cit.
Marx, op cit.
Roncoli (2006), op cit.
54 Hardin, G. (1968). The Tragedy of the Commons. *Science*, 162 (3859), 1243-1248.
55 Weber, J. M., Kopelman, S., Messick, D.M. (2004). A conceptual review of decision making in social dilemmas: Applying a logic of appropriateness. *Personality and Social Psychology Review*, 8(3), 281-307.
56 March, J. G. (1994). *A Primer on Decision Making: How Decisions Happen*. New York: Free Press.
57 Arora, P., Peterson, N., Krantz, D.H., et al. When is a Social Dilemma not a Dilemma? Group Affiliation and Social Goals: Transform Current and Future Payoffs for Cooperation (working paper).
Dawes, R. M. & Messick, D. M. (2000). Social dilemmas. *International Journal of Psychology* 35, 111-116.
Jackson, J.W. (2008). Reactions to Social Dilemmas as a function of group identity, rational calculations, and social context. *Small Group Research*, 39, 673-705.
Krantz, et al. (2008). Individual values and social goals in environmental decision making. In T. Kugler, J. C. Smith, T. Connolly, Y. Son (Eds.), *Decision modeling and behavior in complex and uncertain environments* (pp. 165-198). New York: Springer; Science+Business Media, LLC.
58 Brewer, M. B. & Kramer, R. M. (1986). Choice behavior in social dilemmas: Effects of social identity, group size, and decision framing. *Journal of Personality and Social Psychology*, 50, 543-549.
59 Arora, op cit.
60 City of Knoxville Tennessee. (2009, April 30). Downtown Green Power Initiative Reaches Goal of 400 Blocks Sold. Retrieved from http://www.cityofknoxville.org/Press_Releases/Content/2009/0430e.asp
61 Roncoli (under review), op. cit.
62 Peterson, N., Broad, K., Orlove, B., et al. Participatory processes and climate forecast use: sociocultural context, discussion, and consensus. *Climate and Development*, (forthcoming).
63 Florida spiny lobster transferable trap certificate program. Lobsterconservation.org. Retrieved June 15, 2009, from http://www.lobsterconservation.com/floridalobster/
The Lobster Bulletin. Retrieved June 15, 2009, from http://kodiak.asap.um.maine.edu/lobster/library/publications/bulletin/vol4num2.html
64 Benzion, U., Rapoport, A., & Yagil, J. (1989). Discount rates inferred from decisions: An experimental study. *Management Science*, 35(3), 270-284.
Loewenstein, G. F. (1988). Frames of mind in intertemporal choice. *Management Science*, 34(2), 200-214.
Shelley, M. K. (1993). Outcome signs, questions frames and discount rates. *Management Science*, 39(7), 806-815.
65 Johnson, E.J. and Goldstein D. (2003) Do Defaults Save Lives? *Science* 302, 5649, 1338-1339.
66 Pichert, D., Katsikopoulos, K.V. (2008, March). Green defaults: Information presentation and pro-environmental behavior. *Journal of Environmental Psychology*, 28(1), 63-73.
67 Print Management Information. Rutgers University. Retrieved July 9, 2009, from http://www.nbcs.rutgers.edu/ccf/main/print/transition.php
68 Kageyama, Y. (2009, May 20). Automakers See Green. *Yahoo Finance*. Retrieved from http://finance.yahoo.com/news/Automakers-see-green-apf-15300926.html?.v=5

This book was designed on an Apple Macintosh computer using Adobe Creative Suite software. The display font throughout the book is Rockwell, designed by the staff at the Monotype Type Foundry in 1934. The text font is Joanna, designed by Eric Gill in the 1930s and named for his daughter.

The book was printed with AC Graphics in Hialeah, Florida. The cover paper is FSC, SFI, PEFC, Green-e, and ECF certified; the interior paper stock is SFI and ECF certified. In addition to the print version, this book is also distributed electronically in Adobe PDF format at cred.columbia.edu/guide.

Download this guide and explore additional features at:

CRED.COLUMBIA.EDU/GUIDE

CRED Center for Research on Environmental Decisions

NSF

Charles Evans Hughes Memorial Foundation

Columbia University in the City of New York

Columbia Climate Center, The Earth Institute at Columbia University

Harmony Institute

The Earth Institute Columbia University

The Leonard and Jayne Abess Center for Ecosystem Science and Policy, University of Miami

Rosenstiel School

Lesson 15: What should we do to decarbonize transportation and how can it be done? What part, if any, does hydrogen fuel play moving forward?

NAME: ____________________

Unit Question: Why do our cars run on gasoline and not rocket fuel?

PERIOD: __________ DATE: __________

Lesson 15: What is Power Mapping?

Power mapping is a term used to identify the key players to target when developing campaign strategies. This tool is used to identify a target (person or group), which groups would be against your campaign and how much influence they have, which groups would be supportive of your campaign and how much influence they have, and which groups or people would have little influence.

Power mapping allows a group to identify different relationships and their possible impact to your campaign. By identifying key players and other influential people or groups, your campaign strategy can be tailored to have a larger impact.

The graphic on the following page is used by teams to identify individuals and groups when generating a power map.

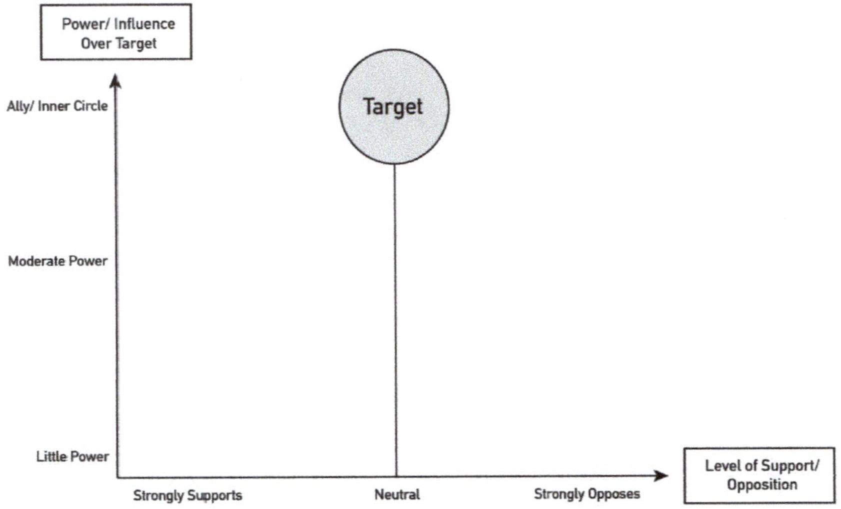

Source: inquiryHub

www.ingramcontent.com/pod-product-compliance
Ingram Content Group UK Ltd.
Pitfield, Milton Keynes, MK11 3LW, UK
UKHW051307061126
16386UKWH00011B/49

9 781792 499333